Project Management in the Real World

Shortcuts to success

Books ·

The British Computer Society

BCS is the leading professional body for the IT industry. With members in over 100 countries, BCS is the professional and learned Society in the field of computers and information systems.

BCS is responsible for setting standards for the IT profession. It is also leading the change in public perception and appreciation of the economic and social importance of professionally managed IT projects and programmes. In this capacity, the Society advises, informs and persuades industry and government on successful IT implementation.

IT is affecting every part of our lives and that is why BCS is determined to promote IT as the profession of the 21st century.

Joining BCS

BCS qualifications, products and services are designed with your career plans in mind. We not only provide essential recognition through professional qualifications but also offer many other useful benefits to our members at every level.

BCS membership demonstrates your commitment to professional development. It helps to set you apart from other IT practitioners and provides industry recognition of your skills and experience. Employers and customers increasingly require proof of professional qualifications and competence. Professional membership confirms your competence and integrity and sets an independent standard that people can trust. Professional Membership (MBCS) is the pathway to Chartered IT Professional (CITP) Status.

www.bcs.org/membership

Further Information

Further information about BCS can be obtained from: BCS, First Floor, Block D, North Star House, North Star Avenue, Swindon, SN2 1FA, UK.

Telephone: 0845 300 4417 (UK only) or + 44 (0)1793 417 424 (overseas)

Email: customerservice@hq.bcs.org.uk

Web: www.bcs.org

Project Management in the Real World

Shortcuts to success

Elizabeth Harrin

The British Computer Society
Publishing and Information Products
First Floor, Block D
North Star House
North Star Avenue
Swindon
SN2 1FA
UK

www.bcs.org

ISBN 1-234567-89-0
ISBN13 978-1-234567-89-0

British Cataloguing in Publication Data.
A CIP catalogue record for this book is available at the British Library.

captured, authored, published, delivered and managed in XML
CAPDM Limited, Edinburgh, Scotland **www.capdm.com**

Contents

List of Figures and Tables

About the Author

Elizabeth Harrin has worked within the financial services industry since 1998. Now a senior project manager for the global financial services company AXA, she has successfully led a wide range of technology and business projects, involving managing international project teams across multiple sites. Elizabeth is a PRINCE2 Practitioner and is trained in the Six Sigma process improvement methodology as a Black Belt. An alumnus of the universities of York and Roehampton, Elizabeth is a keen gardener, a hobby which is seriously hampered by the fact that she currently lives and works in Paris, where vegetable patches are non-existent.

Acknowledgements

I am indebted to the many managers and companies who have generously given their time and offered their experience for these case studies.

In many ways, producing this book was a family affair. Numerous errors were spotted and erased by the eagle eyes of Pauline Harrin. Several of the diagrams were produced by Caroline Harrin, whose talent for turning my scribbles into graphics that mean something never fails to amaze. And if it wasn't for my father Alan, who taught me how to program on a tape-driven computer, I might not be working with technology today.

Thanks are due in particular to my husband, Jon Borley, who was generous with his great ideas, suggestions and cups of tea.

I am also grateful to the team at BCS, in particular Matthew Flynn for his patience and support throughout the process. His input, plus that of the two anonymous reviewers, helped me improve the text. The book has been extensively proofread and reviewed, and so any errors or omissions in it are strictly my own.

Glossary

Actual cost of work performed (ACWP) Amount of money spent on the project activities up until a given date.

Assumptions Statements made during a project that are not based on known or certain facts.

Baseline Stake-in-the-sand view of a project schedule, budget or other moveable activity that provides a comparison of the actual situation against the expected situation.

Business-as-usual Day-to-day activity as distinct from project activity.

Change control Process of managing change in a controlled way.

Change management See *change control.*

Contingency Provision made within the project planning stages to allow for unforeseen circumstances; usually built into the budget or schedule.

Critical path Longest route through a project plan; collective name for the group of tasks that must be completed on time in order for the project to deliver to the planned end date.

Critical path analysis Process of establishing the critical path; can include drawing out the critical path diagrammatically.

Deliverable Something tangible delivered as a result of the project.

Dependency Relationship that links the order in which activities are carried out. Task B is said to be dependent on task A if the start or finish date of task A must be reached before task B can start.

Earned value analysis (EVA) Method to establish the budget and schedule position of a project based on resource planning.

Estimate at completion (EAC) Total budget required to finish the project, calculated by adding together estimate to complete and expenditure incurred to date.

Estimate to complete (ETC) Budget required to finish the project calculated from a given date to the project end.

Ice-breaker Activity or short game used to introduce team members to one another; used in workshops, long meetings and at the beginning of projects.

Issue Risk that has actually occurred or another known circumstance that may impact the project's outcomes.

Issue log Document listing all the issues that are impacting the project; updated with the activities required to actively manage and resolve each issue.

Issue register See *issue log.*

Milestone Date by when a particular chunk of work is due to be completed.

Network diagram Visual representation of a project plan, showing the links between each task; used in critical path analysis.

Plan Document, or several documents, detailing exactly what the project needs to do in order to deliver the objectives; a practical analysis of what deliverable will be produced by whom and when.

Pilot phase/stage Project implementation in miniature to test and assess the impact of the deliverables before the project is fully rolled out.

Programme Collection of projects with a common theme, sponsor or reporting process.

Proof of concept Test of the project deliverables in a controlled environment; shorter and more laboratory-based than a pilot.

Post-implementation review See *post-project review.*

Post-project review Meeting to evaluate the project's successes and challenges and record any learning for future projects; a way of sharing corporate knowledge.

Project board See *steering group.*

Requirements document Document that records all the things (requirements) the end user wants from the project; used as a basis for technical documentation.

Risk Statement of the possibility that something unforeseen will happen to the project that will have a negative or positive impact on the outcome.

Risk log Document listing all the risks that may impact the project; updated with the activities required to minimize each risk.

Risk register See *risk log.*

Risk response Approach to managing a risk; typically one of: avoidance, transference, reduction, acceptance.

Schedule Document listing all the tasks that need to be done in order to complete the project and the dependencies between them; the project calendar.

Scope statement Description of what is included in the project and what is not; covers deliverables but also groups of people impacted and the reach of the intended activity.

Sponsor Senior manager who heads up the project; person who champions the work and to whom the project manager reports with project progress.

Stakeholder analysis Exercise to determine the interest and influence of stakeholders to establish their support for the project and what can be done to influence their position.

Stakeholder mapping See *stakeholder analysis.*

Stakeholders People who have an impact on, or who are interested in, the project.

Steering committee See *steering group.*

Steering group Group made up of the project sponsor, project manager and one or two other key stakeholders; this group is responsible for decision making.

Success criteria Standards by which the project will be judged at the end to decide whether it has been successful in the eyes of the stakeholders.

Test scripts Documents explaining the step-by-step method required to test a deliverable; given to testers to ensure testing is done in a methodical way.

Workstream Part of the project that can be managed as a discrete chunk; led by a workstream leader.

Preface

Project Management in the Real World won't teach you how to be a project manager. It's not going to show you how to set up your first project, walk you through it and see you out the other end with all the benefits realized.

Traditional project management books do that, following the project life-cycle with chapters on project definition, initiation, execution, closure and so on. This book is different.

It's for people who already know that a project has a beginning, a middle and an end and who want to take project management further. It's for people who know the theory and feel there must be an easier way to get things done. It's over 250 years' worth of combined project management experience distilled into 200 pages so you can see how other people run their projects outside the management texts and research papers: how projects get done in the real world.

This book is organized into five sections: managing the project budgets, scope, teams, plans and yourself as project manager. Wherever you are in your project, you should be able to easily find information relevant to the particular situation you find yourself in.

Each section is divided into short chapters, which explore discrete elements of the business of project management. Each chapter includes an anecdote from a manager who has been there and done it or a case study from a project with a valuable lesson to be learnt. For clarity, and also because this book is designed for people without much time to study project management theory extensively, each short chapter covers one discrete point that you can put into practice immediately: you'll understand both why and how things can be done. Dip into the chapters at random and pick a section, or make your way methodically through the section most relevant to where you are in your project at the moment. If a topic particularly grabs you, flick through the further reading suggestions and references to find ways to take it further.

Throughout the book, you will see icons in the margins to guide you to important information in the text. Here's the key:

HINT

A hint or tip to help you apply the knowledge in the chapter.

ANECDOTE

An anecdote or case study: real-life experiences from project managers who have been there.

GOLDEN RULES

The golden rule to remember, even if you don't remember anything else about the chapter.

DEFINITION

A definition of a project management term or principle.

WARNING

A potential trouble spot or project management pitfall.

Some names and project settings have been changed or disguised at the request of interviewees. The chapters cover the elements that I feel are most relevant to modern project management but are frequently overlooked. It has not been possible to include everything I wanted, and I'm sure you'll have a favourite hint, tip or memory that you believe other project managers could learn from. Please email me with your ideas for another volume at elizabeth@elizabeth-harrin.co.uk.

Elizabeth Harrin
Paris, July 2006

Foreword

In this book, Lonnie Pacelli is quoted as saying 'Surprises are for birthdays'. It was in fact a few days after my birthday that Elizabeth Harrin approached me to write this foreword, a very pleasant surprise!

In the media, we frequently read or hear about project failures, which consequently adversely affects the reputation of all project managers. The success stories rarely hit the headlines. As Chair of the BCS Project Management Specialist Group (BCS PROMS-G), I am, *inter alia*, responsible for promoting professionalism in our specialist sector of the information technology (IT) industry. By providing, through our countrywide events, timely and relevant information on industry developments and by sharing lessons learned, PROMS-G promotes continuing professional development. Our aim is that project managers, and therefore their projects, will be increasingly successful and hit the headlines for the right reasons. Elizabeth is one of our 5,000-plus valued members and an occasional speaker.

A key skill required of all project managers is to identify potential risks and to remove or mitigate their effect before they become issues. While we all appreciate pleasant surprises, it is the unpleasant ones that have the most adverse effects on a project. Regardless of whether you are an experienced project manager, it is highly probable that you will come across both types of surprise.

It is, however, impossible for project managers to foresee all situations that may arise. While we should all attempt to continually develop our professionalism and to keep abreast of developments in our own industry or particular area of expertise, this may not always be possible due to the large amount of change that occurs. It is therefore imperative that project managers are able to focus on these changes and assess their impact rather than spend their precious time resolving underlying project management issues such as budgets, processes and so on.

This book aims to assist with getting the latter right. It is a valuable reference point for ensuring that a project has the underlying essential processes and authorities in place and that they are working as intended. Some of the pitfalls that await the unwary or unskilled are identified and guidance is provided on how to avoid them. In following these recommendations, and not spending time resolving basic issues, a project manager's time will increase, allowing him or her to focus instead on the more critical risks and issues.

It is no surprise to me that Elizabeth has written a book that is very easy to read and that you can dip in and out of as required. Each part is self-contained and will provide that nugget of information you have been looking for. Elizabeth has collected the issues, anecdotes and success stories not of entire projects but of the elements within them. I am pleased that so many project managers were willing to share their experiences, because it is only by sharing and learning from these experiences that we can all continually develop and enable our professionalism to grow. All project managers, whether working in IT or in other industries, will identify easily with the lessons learned. If you find something works for you, then please pass it on. By the way, PROMS-G is always looking for speakers for our events.

The phrase 'Surprises are for birthdays' is one of the mantras that should guide us in all aspects of project management. As a professional project manager and chair of PROMS-G, perhaps I should have anticipated the pleasant surprise of being asked to write this foreword. On the whole though, I would rather focus on avoiding the unpleasant surprises and leave the pleasant surprises just as they are. Elizabeth's book helps to do just that.

Ruth Pullen
Chair, Project Management Specialist Group
British Computer Society
www.proms-g.bcs.org/
March 2006

Section 1
Managing project budgets

Know that with a farm, as with a man, however productive it may be, if it has the spending habit, not much will be left over.

Marcus Porcius Cato (BC 234–149), *De Agricultura*

More than one-third of projects have a budget of over £1 million, and so knowing how to handle the finances is an essential part of a project manager's repertoire. The initial budget is often just a starting point. An incredible 56 per cent of projects are affected by budget changes, and that's not just a one-off financial revision. The average project, if there is such a thing, has its budget revised 3.4 times.[1]

Keeping on top of all this is not always easy, and it is made harder by the fact that project managers themselves don't always get control over the money. This section covers how to manage project variables over which you do not necessarily have authority, how to find out who has that authority, and how to manage the relationship with the budget holder. Many projects do not appear to have budgets at all, and Chapter 11 looks at working effectively in that environment. This section also looks at reporting, tolerances and contingency.

1 Create a realistic budget

Even the smallest project will have overheads, your time as the project manager being a minimum. Nearly all projects will have more than that, so part of your role in setting up the project is to define and propose a budget for the work and get that approved.

a

BRAINSTORMING THE BUDGET

'I haven't had much experience handling money, so doing my first project budget was really hard,' says Emily Jones, a junior project manager in a small public relations consultancy. The project was to revamp a room that had been used for storing spare furniture into a new area for holding workshops. 'My sponsor left me to it, so I had to work out the money I thought I'd need by myself.' Jones set up a brainstorming session with her team and asked them to help her identify all the likely costs for the project. 'We came up with the obvious ones like staff salaries and buying the new office furniture really quickly,' she says. 'Then I asked them to be more creative, and someone said "Hiring a projector for the staff briefing." OK, so that might not sound really creative, but as our company projector had just broken, and we were scheduled to do a presentation on the project in three weeks at a briefing for all 45 staff, it was a cost I certainly hadn't thought of.' In fact, Jones hadn't even known the company projector was broken. The replacement was on order but not due to arrive for another five weeks. Jones wanted her presentation at the company briefing to be professional, and projector hire was not a great deal of money, so a member of the team was tasked with finding an estimate and the cost was added to the budget. 'On the subject of hire, we also came up with hiring a van to take the old furniture to a charity warehouse. We could have had the council take it away for free, but we decided we'd rather it went to a good cause, so that cost ended up in the budget too.'

Jones split the identified costs into groups. 'In the end we had a group of charges for manpower for our time and one part-time contractor, and a group of charges for putting in a new telephone, the decorating costs and some miscellaneous things. I added a contingency line of 15 per cent of the overall budget as I knew many of the costs were just estimates,' Jones continues. 'I explained to my sponsor that this was for risk management and he cut it to 10 per cent. I thought that was reasonable, and he approved the budget on that basis.'

Creating a budget is like putting together a project schedule, which we'll look at later. You can work out how much money you will be spending based on what you know needs to be done, just as you work out how much time the project will take based on the same information. Think of the budget as a shopping list of all the things you need to buy to make sure the project gets completed. Just like a trip to the supermarket, you might not end up spending exactly what you expected but at least the list gives you a reasonably accurate starting point. 'When planning, assume your budget will not be increased or decreased during the project,' writes George Doss in the *IS Project Management Handbook*. 'Budget changes . . . are adjusted through negotiations with the project sponsor based on circumstances at the time.'[2]

There are five steps to creating a project budget:

(i) Identify the resources required for the project.
(ii) Estimate the cost for each of those resources.
(iii) Document the costs and calculate the overall figure.
(iv) Submit the budget to your steering committee or sponsor for approval.
(v) Find out your budget code.

Let's take each of those steps in turn.

IDENTIFY THE RESOURCES REQUIRED FOR THE PROJECT

Review the schedule, project initiation document and any other documents you have to identify the activities that need to be completed. Draw on your stakeholders and project team to brainstorm anything else that might be required, e.g. travel, accommodation, couriers, equipment. Will your project have to pick up the costs incurred by other areas of the business that are impacted by the work you are doing? Ask other managers who have done similar projects to validate your list.

ESTIMATE THE COST FOR EACH OF THOSE RESOURCES

Every step, every task of the project will have associated costs. Projects that do not have full-time staff may avoid paying for the entire salary of anyone working on it, so ask the finance department whether there is a list of standard chargeable rates per 'type' of employee. For example, your project might have to pay £1,000 per day for an expert manager but £650 per day for a junior marketing executive. Some of these costs may be just 'paper' prices – especially for internal resources. They are just figures you plug into the business case, but in reality money never changes hands. Check out your company's rules for charging for business resources, and also check with each department head about their expectations. For example, if they are loaning you a person, then they may expect the project to fund a temporary resource to backfill that person's day job.

A NOTE ON ESTIMATING

Given the flexible nature of budgets, and projects in general, it's very hard to pin down costs to an exact figure at this early stage of the project. And it's not a good idea either, unless you are absolutely 100 per cent sure that your estimation is spot on and will not change.

At this stage, present your estimates as a range rather than a fixed sum. This means your overall project budget, once you have added up all your estimates, will be between £*x* and £*y*. It's this range that you present to your steering group and sponsor.

Presenting a range means a little more flexibility later on. It also gives you the chance to start managing the expectations of your sponsor now – they will have to come to terms with vagaries and changes as the project progresses, so now is a good time to start explaining the nature of project management.

DOCUMENT THE COSTS AND CALCULATE THE OVERALL ESTIMATE

Companies that carry out a lot of projects will probably have a standard template for submitting a budget, so find out whether a form already exists. Create your own form in the absence of anything standard, using a method that suits you, for example a computer spreadsheet. The advantage with an electronic budget spreadsheet over using a word-processing package or a paper system is that the figures will update automatically, reducing the risk of manual error and saving time. Group together similar costs, so you have subtotals as well as an overall total, and include a line of contingency for risk management. Compare your budget range with any amount given to you by the project sponsor, and see below for what to do if the figures don't match.

SUBMIT THE BUDGET TO YOUR STEERING COMMITTEE OR SPONSOR FOR APPROVAL

Once you have your budget written down, it needs to be approved before the project can continue. Your sponsor or steering committee is the first point of approval. They will advise you on whether the budget needs another level of approval from finance, a central planning committee, an IT authorization forum or another group, depending on where the funds are actually coming from.

WARNING

More often than not, you'll be asked to kick off the project without budget authorization. In the real world, there are deadlines to meet that won't wait just because the budget committee meets only on the last Tuesday of the month. If you're asked to start work without the relevant approvals, get on with it! But make sure you have something in writing to cover yourself against any expenditure incurred during the time you're working without an approved budget.

FIND OUT YOUR BUDGET CODE

Assuming all goes well, the budget will be approved and you will be given the go ahead to spend the money required. Any expenditure needs to be tracked back to the project so the budget holder can keep an eye on what is being spent. The project might be allocated its own pot of money, ring-fenced from other budgets, in which case you will probably have a cost centre code of your own. Alternatively, the project might be allocated a portion of the budget for a particular department. If this is the case, ask your sponsor how they want you to identify project spending. A non-committal answer means you will have to invent your own code, perhaps the project number or a shortened version of its name. When you sign an invoice or raise a purchase order, use the code to ensure the expenditure can be tracked back to the project; make certain that anyone else who has the authority to use the budget does this as well.

WHAT IF MY SPONSOR ALREADY HAS A BUDGET IN MIND?

Just because this is a sensible five-step approach that allows you to analyse the work involved and cost it accurately does not mean it is followed by all project sponsors. For many reasons, you could find yourself working on a project where the sponsor already has a set figure in mind. Some sponsors will knock off 10 per cent from your total because they believe the numbers are padded. Others may be compelled to halve the budget because someone higher up the chain expects cuts across the board.

If you put your mind to it, you can complete any project to a specified budget – at a hidden cost. Corners will need to be cut, quality might suffer and the customers may not get everything they thought they would. Present your steering group with a couple of options for reducing your proposed budget to their predefined figure, making the trade-off between quality, time, scope and cost. They may still tell you that it's their budget you need to follow, but at least you have explained the risks of delivering to a certain abstract budget figure and you have your planning documentation to back up your arguments.

GOLDEN RULES

To create a realistic budget, base your predicted expenditure on your project planning documentation and get the budget approved as quickly as possible to prevent any delay in starting work.

2 Calculate the true cost

The cost of your project is probably not as transparent as you originally thought. Digging into the detail will help you really understand how much the project will cost and, therefore, let you avoid any nasty surprises later.

a

MILLENNIUM MONEY MADNESS

The Millennium Dome opened its doors in Greenwich, London, on New Year's Eve 1999 and managed to keep them open throughout 2000. The project delivered on time and proved a memorable day out for many of the 5.5 million paying visitors and another million who were entitled to free entry. Despite this being more than double the amount of paid-for entries to any UK visitor attraction in the previous year, the Dome still found itself in financial difficulties and will be remembered for its vast overspend.

At the end of 2000, the National Audit Office produced a report evaluating the financial problems of the project. Two things stood out: the overly ambitious visitor targets – the New Millennium Experience Company that operated the site originally planned for 12 million guests – and the fact that running the project was complicated by both not having adequate financial management systems in place and the complex relationships between the various stakeholders. This made it difficult for the company to forecast the true cost of the project and resulted in a budget shortfall. The Dome found itself requiring more funding from the National Lottery. Based on the experience of the Dome, the National Audit Office recommended that future projects should 'proceed on the basis of a full understanding of the cradle to grave costs'.[3]

DEFINITION

Shim and Siegel define a budget as 'the formal expression of plans, goals, and objectives of management that covers all aspects of operations for a designated time period.'[4]

There are two important elements here. The first is 'all aspects'. Figure 2.1 shows the two categories of expenditure you should consider in detail:

- Project management costs: the costs of doing the business of project management.
- Project deliverable costs: expenditure related directly to what the project is going to deliver.

Source: State of Western Australia

FIGURE 2.1 *Types of project expenditure*

When you are analysing your budget to guarantee it includes 'all aspects', using these two categories will help clarify your thinking.

You might only be equipping a new office and not putting on a once-in-a-millennium style exhibition, but the lessons from the Dome's experience are worth implementing nevertheless. When you are trying to calculate the total cost of your project, brainstorm all the things that will cost money in both categories. Include paying for your team members (project management costs), buying software and consultancy, and funding anything that will change as a result of your work, such as new stationery and user guides (project deliverable costs). Maybe you will have to provide training courses (project management costs), which will include the costs of a trainer, room hire, refreshments, delegate transport and accommodation, host large meetings off site (project management costs) or pay for documents to be translated (project deliverable costs).

Once you have a comprehensive list, add to it all the things that you believe will not cost anything: the business users' time for testing, your time and so on. The point of doing this is to have a documented list of 'free' things. These form project assumptions and will be validated as the project progresses. Your sponsor should verify this list, which can be included in the project initiation document. If at any point you find that you were wrong and that you do have to pay for items you believed were free, you can explain to your sponsor that the budget increase is due to these assumptions being incorrect. We'll see more about assumptions in Chapter 22.

Once you've worked out all the elements and likely costs, decide on how to report your budget. If your sponsor is interested only in cash out the door, it will not be necessary to report how many days the 'free' human resources have spent working on the project.

The second important phrase in Shim and Siegel's definition is 'for a designated time period'. The project does not necessarily end as soon as you have delivered whatever it was you set out to deliver. There could be (and are likely to be) costs incurred in the final stages, the 'grave' part of 'cradle to grave'. The budget should include adequate provision for any end-of-project expenditure. That means, for example, the charges for decommissioning any now-defunct system, product or literature, retraining for staff who now don't have the right skills, and generally making sure the old status quo is not someone else's financial headache.

Time periods have another important impact on budget management. If your project stretches over two financial years, you will have to apportion your budget appropriately and you might have to navigate your way through the maze of year-end accounting and the accrual process. Get some advice from an old hand if you're facing this for the first time, as the rules differ from one organization to another – although it might take you a little while to find someone who can explain them clearly!

GOLDEN RULES

Don't guess what your budget is supposed to pay for. Do your own research to understand fully all the explicit and hidden charges to help you control costs more accurately over your project's lifecycle.

3 Agree a budget tolerance

DEFINITION

A budget tolerance is a range within which you can spend without having to report back to your sponsor and ask for more money.

Budget tolerance is particularly useful at the end of a project as you near the delivery date. If you have a budget of £80,000 with a tolerance of 10 per cent and you complete the project for £85,000, you have still delivered within the parameters set by your sponsor. A budget tolerance of 10 per cent means you can deliver the project 10 per cent over cost without having to get special permission to do so.

NOT A PENNY MORE . . .

Peter McDonald, an engineer working in Wales, thinks back to his first project: 'It was quite small actually,' he explains. 'I was just starting out in project management and was working in a team improving the process for getting car parts off a distribution line more quickly. The project budget for non-resource spend was small and, as no one else wanted to do it, I got put in charge of monitoring the expenditure.' The project manager delegated the responsibility for tracking the budget and ensuring the team did not spend more than had been agreed to McDonald. 'I was really nervous and I watched every penny,' he adds. 'I suppose it was about £30,000, which, considering what I manage now, really wasn't that much, but at the time on my just-out-of-university salary it was massive.'

McDonald's team had eight months to analyse the existing process, come up with a new one and implement the changes successfully. The analysis went well and within three months the team had got agreement from the factory management to implement their new process. There was no budget for buying new machines, so the changes were subtle but effective. 'We ended up by streamlining the process in the warehouse,' McDonald says. 'We couldn't make changes to the actual manufacturing part of the process as it was prohibitively expensive, but we cut out some of the admin steps.'

It was re-engineering the paperwork that used up most of the budget. The project manager consulted with McDonald and purchased a system for handheld scanning machines to remove the need for manual checking when boxes of the car parts were ready to be shipped out. 'The technology seems antiquated now, but it was revolutionary for us,' McDonald says. 'But the problem came when the invoice dropped on to my desk.' He had forgotten to add the cost of delivery charges and the three-year warranty the company had purchased. With those additional amounts, the budget was now running at 3 per cent over. McDonald started to worry. 'There was no way I could pull the budget back in line, especially as I wasn't the project manager,' he explains. 'So I had to confess.'

McDonald took the project manager to one side and informed him of the mistake. 'He asked me if we were still on track to deliver everything else within budget, and I said yes. Then he told me not to worry as he had agreed a 5 per cent tolerance with the sponsor.' McDonald was relieved but annoyed. 'I should have been told that at the beginning when I was given the responsibility, but I didn't ask either,' he says. 'Since then, I've made sure I know what the tolerance levels are for my projects so I'm aware if there is some degree of flexibility.'

At the beginning of the project, discuss a budget tolerance with your sponsor. This is a way of minimizing effort for them, as you will not be bothering them with frequent change requests for tiny budget increases. Agree an appropriate tolerance and write it into the project documents. What is appropriate will depend on the size of the project, the size of the organization and the organization's maturity with regard to projects. The tolerance will not be 'used' until the end of the project, but it will help you monitor performance and track how you are doing compared with your initial estimates. As soon as the project looks like it will fail to deliver inside the tolerance, you know you have a problem to address. Tolerances can be used like early warning systems: they give you a little bit of leeway but enable you to tell quickly how far you are from your targets if the project begins to stray off course.

HOW IS CONTINGENCY DIFFERENT FROM TOLERANCE?

A contingency fund is an amount of money set aside for project emergencies. It is a project's overdraft. The project manager needs permission and a good reason to spend it, but it is assumed that it will be used at some point or other through the project. Contingency can be for any amount – sometimes even 50 per cent of the original project budget. The project manager calculates an appropriate amount based on the project's risk factors and negotiates the final allocation with the sponsor.

Budget tolerance is the amount by which the project can be delivered over (or under) budget without anyone being concerned. It's usually a small amount represented as a percentage. Tolerance is calculated either as a straight percentage of the core budget estimate or as a percentage of the core estimate plus the contingency fund. As you should assume the contingency will be spent, it's better to agree a tolerance based on the latter.

The amount of tolerance is set by the sponsor or main budget holder, based on your recommendation. It's an acceptance of the fact that you might need a little extra and that in the grand scheme of things, the overall company budgets can handle a little flexibility.

HINT

This type of tolerance relies on your sponsor agreeing to a degree of flexibility within the project budget. But what happens if they say no? If you are not allowed an explicit tolerance, the pressure is on to deliver on budget. A contingency fund becomes even more useful. But what if you can't get your sponsor to agree to one of those either? Consider padding your budget estimates a little so you give yourself a cushion of implicit contingency. It's sneaky, but it will give you more flexibility with the finances later if you can get away with it.

Budget and time tolerances are often set together as part of the same conversation with a sponsor, which means they can be plotted graphically as in Figure 3.1. This graph shows that the sponsor is happy for Project Whirlwind to finish between mid-September and mid-October and cost between £71,250 and £78,750, although the target is to finish at the end of September and spend £75,000. The sponsor and project manager have agreed a budget tolerance of ±5 per cent.

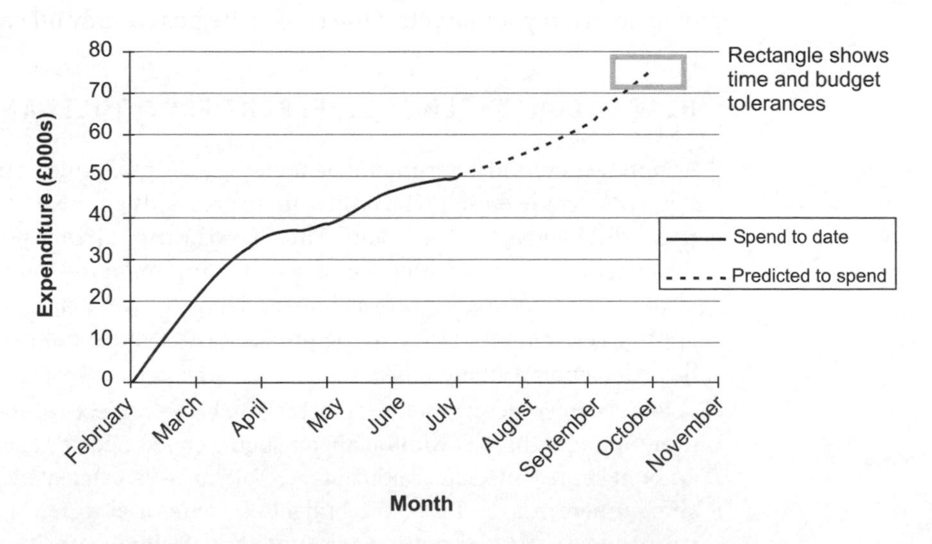

FIGURE 3.1 *Time and budget tolerances for a hypothetical project*

Minus tolerances are important too. If you bring a £75,000-project in on time and to the required specification but spend only £20,000, someone will start asking questions. Tying up money in project budgets that could be put to better use elsewhere is not good business practice and will cause significant issues in small organizations. If you believe your budget will not be spent within the tolerance levels, study why, double-check and then raise a formal change to make your sponsor aware of the issue and adjust the budget according to your new calculations.

GOLDEN RULES

Agree a budget tolerance early on with your sponsor, even if you won't come to use it until later in the project.

4 Track estimate to complete

It is easy to track how much of your precious project budget you have already spent, assuming you keep copies of invoices and timesheets. It is easy to assume that you will just use up the rest of the available money evenly between now and the end of the work. This approach gives you a false impression of how you will spend the rest of your budget. For example, you could be 50 per cent of the way through the project and have spent 50 per cent of the budget. However, if a big purchase has not yet happened, having only half the budget left could be a sign of a trend towards overspending. Project costs are rarely distributed evenly throughout a project: some projects spend very little early on and then incur all the costs in the last few weeks.

DEFINITION

Estimate to complete (ETC) is the amount of money you predict will be required to finish the project. Tracking ETC gives you an accurate view of the projected budget you need to get everything done.

REAL-TIME BUDGETING

'The most important things I reported on projects were budget to date, actual to date and estimate to complete,' says Lonnie Pacelli, who worked at Accenture before setting up his own consultancy. His 20 years of experience at delivering projects and his previous role as Microsoft's director of corporate procurement overseeing $6 billion of expenditure have given him a clear insight into how to handle project budgets.

'One of the most overlooked components of budget management is a realistic estimate to complete,' he continues. 'Too many times a project manager will just subtract their actual to date figure from the total budget to calculate the estimate to complete.' Pacelli believes this gives an unrealistic view of the money needed to finish the work. He suggests a better route is to determine the ETC based upon the tasks the project manager knows are still to be done, and not on a vague hope that the project will come in on budget.

'A realistic view of estimate to complete is a major sniper in the weeds on projects. Too many project managers assume that everything will go perfectly for the remainder of their project, even if the project to date has been difficult,' adds the Washington-based president of Leading on the Edge International, a management consulting and self-study leadership education firm. 'Don't just say "I have $1 million total budget and have spent $600 thousand, so my estimate to complete is $400 thousand." Look at the remaining amount of work left to complete and realistically cost out the work.'

This approach is more robust and will also allow you to highlight potential budget issues early. 'I reported budget status to the executive sponsor, steering committee and also the project team,' Paccelli says. His advice for project managers who find themselves reporting that their projects will go over budget on completion is to use a realistic ETC figure and to ask for more money just once. 'Going back to the well more than once erodes management's confidence in the project manager and creates doubt as to whether or not there is another surprise waiting around the corner,' he explains. He cautions against holding back the information about the cost rise and is clear that raising the issue early is a safer tactic. 'Don't assume some wonderful thing is going to happen which will cure all of your budget ills,' he says. 'Surprises are for birthdays, not for project budget management.'

HINT

Calculating the ETC for your project is really straightforward. You already know how much of the work is done, so you know how much is left to do, based on the latest version of your plan. You can work out how much the still-to-do work will cost, based on your budget assumptions. That figure is your ETC. It is more useful to express this figure as a financial amount rather than a number of hours or days of effort to be sure the ETC takes non-resource costs into account. Also bear in mind that you may not have all the invoices from third parties. Their charging patterns may mean they invoice a month in arrears, after the work is done. You could find yourself in a situation where the work is complete but not yet charged for. Just ask them for the relevant figures if you're not sure.

The ETC plus the figure you have already spent represents the budget you expect to have spent at the end of the project.

DEFINITION

The budget already spent is known as actual cost of work performed (ACWP). Your estimate of the total amount spent on the day you close the project is known as estimate at completion (EAC), and this is the figure that will interest your sponsor. This calculation is summarized in Figure 4.1.

ETC + ACWP = EAC

FIGURE 4.1 *Calculating estimate at completion*

Keeping track of EAC is a simple way to predict budget overspend. It provides an early-warning mechanism and allows you to plan how to tackle an increasing budget. In fact, if you are doing a technology project, it is highly possible that your monitoring will show escalating costs. Between 30 and 40 per cent of IT projects fail to stick to their original budget.[5] Studies show it is worse in other industries: 90 per cent of transport infrastructure projects overspend.[6]

The reasons for overspending are many and varied, but the costs to be most aware of are those associated with your human resources: 'Generally, labor costs overrun significantly more than can satisfactorily or easily be explained,' report Eden *et al.*[7] 'Non-labor costs can more usually be tied to specific causes.' Fortunately, using ETC and EAC to track expenditure on your project can help you identify any trends towards extra costs at an early stage.

ETC, ACWP and EAC are components of a larger financial tracking method called earned value analysis (EVA). EVA is a tool that shows whether you are over or under budget, and behind or ahead of schedule, at any given moment in the project.

There is not the space here to explain EVA in its entirety. If you are just starting out in project management and have a relatively small budget, calculating ETC and EAC is a good solid start. As EVA takes time and effort to do properly, it adds limited value to small projects. With a larger project, you may find the EVA method useful to help you understand where you are. Portny[8] has an excellent appendix describing EVA and including worked examples, so if you are interested in taking these measurements further, try that as a basic introduction.

WARNING

Remember: any financial or mathematical calculation can offer only a numerical representation of a project's situation with regard to the overall costs and schedule. No figures like this will ever give you a narrative explanation of why your project has ended up here or why it appears to be going off the rails – you will have to work that one out for yourself.

GOLDEN RULES

Tracking ETC and EAC will give you an early warning of possible project overspend and real-time useful information to

report to your sponsor, but you will need to add the explanation of why yourself.

5 Have a contingency fund

DEFINITION

A contingency fund is money set aside at the start of a project to be used in case of need, for example to offset unforeseen increases in costs. The amount of this ring-fenced budget depends on the level of risk the project faces and also on the overall project budget itself.

ANECDOTE

Hanford, Washington, is home to one of America's largest nuclear storage plants run by the United States Department of Energy (DOE). There are 177 waste tanks on site, storing more than 55 million gallons of high-level radioactive waste underground – that is equivalent to an area the size of a football field over 150 feet deep. The DOE launched an 11-year project in 2000 to build facilities at Hanford to treat and prepare the waste for disposal. Around the same time, the DOE launched a project management initiative designed to counteract the department's poor record of inadequate management of contractors. The initiative recommended that contingency funding be built into a project budget according to the project's degree of risk. Unfortunately, at the time of signing a contract with the construction company in December 2000, the project management initiative had not been implemented fully. When an internal DOE assessment was carried out, it became clear that the department had signed a contract with a flaw: the cost baseline of $3.97 billion was so low that the project had only a 50 per cent chance of delivering against it.

The DOE took steps to address the gap in April 2003 and revised the cost baseline to include a $550 million contingency budget. The DOE also set up a governance panel consisting of both DOE and contractor personnel to manage the additional funding and to monitor spending. The aim of the contingency budget was to counter unforeseen cost increases across the life of the project. The team also allocated an additional $100 million to be used to mitigate unforeseen technical and management risks.[9]

The project is due to complete in July 2011 at an estimated cost of $5.7 billion, although it remains to be seen whether that figure is the final prediction. A May 2004 study by the US Army Corps of Engineers reported that the project would probably need another $720 million, and DOE officials have confessed that they did not know whether the project was meeting its performance baselines. However, an audit in March 2005[10] highlighted that project reports were still showing that the cleanup work was on target to meet the approved baseline of $5.78 billion.

At the beginning of your project, you will need to calculate the expected budget required in order to deliver the work. This budget figure is the first step to being able to work out a reasonable contingency. The project risk register will also be important, as your knowledge of the risks inherent in the delivery will inform your decision about how much contingency is required: the riskier the project, the higher the amount of contingency budget. Calculating contingency is not really a science. Once you have a full understanding of the work the sponsor expects you to deliver, you must take a best guess at the figure and ask them to approve it. Your company might use a formula for calculating contingency, but as it depends so inherently on the risk factors for each individual project, it is hard to give a one-size-fits-all equation. A project that has been run several times before, with experienced staff and solid estimates for both time and expenditure, will need little, if any, contingency. A project using new technologies or doing something that the company has never done before will require a large contingency fund to offset any unforeseen costs.

BUT I HAVE NO IDEA WHAT 'REASONABLE' CONTINGENCY IS

It is not a good tactic to avoid setting a contingency budget because it's too hard to work out what is reasonable. Add a contingency line to your budget of 10 per cent. This at least gives you a starting figure to begin negotiations with your sponsor, and it will give you some leeway if it does turn out that your estimating has been a little wayward. The more experience you have with budgeting, and with projects, the easier it will be to predict what amount would make a reasonable contingency fund.

Whatever your chosen figure for the contingency budget, you will have to convince your sponsor you need this allocation. Explain that it will cover things such as cancellation costs for training courses when your delegates are ill, unforeseen taxi bills (small items soon mount up) and value-added tax (VAT) on a supplier's quote that you took to be inclusive but was actually extra. It may help your case to explain to your sponsor that the money will be kept separate from the main budget. It will not be physically in a separate bank account, but it is a separate line in your budget tracking. Also explain the situations in which you will call upon those funds and the approval process required to spend them. You can elucidate further that having the additional money set aside now will make dealing with any project changes or unforeseen events much less painful in the future – and if you do not use the budget, it will always be there for another project.

It's important to explain at this point that the contingency fund is there to help cover costs for work that is required to meet the original objectives. If the sponsor wants changes to the scope, they should be costed and their impact analysed separately. The sponsor should find additional funds to pay for changes. For more on managing changes, turn to Chapter 13.

Ruthanne Schulte from Welcom outlines four steps to managing budget contingency:[11]

(i) Calculate the amount of contingency budget for each short phase or unit of work, not at a project level.

(ii) Hold the money separately and get management approval to move it across to the main budget.

(iii) Once management have approved the spending, increase your budget appropriately so you have an accurate idea of how much you have spent or are predicting to spend.

(iv) If contingency is not required, give it 'back' to the company so it is not included in any profit calculations at the end of the project.

The important point here is to make sure that you report the use of any contingency funds transparently. It will give your sponsor more confidence that you are acting in an accountable way and should also dissuade you from overspending irresponsibly.

GOLDEN RULES

A contingency budget allows you to react more quickly to any unforeseen events that plague your project. If you are able to convince your sponsor to let you have a contingency budget, set the amount based on project risk and ensure you have a mechanism by which to authorize and monitor the use of the funds.

6 Gain buy-in for collective responsibility

A budget is the responsibility of everyone in the project team, regardless of whether the team members individually have anything to do with the administration or invoice handling on a day-to-day basis. Every member of the team incurs costs (for their time as a minimum) and therefore should appreciate their role in making sure the project stays within budget.

a

MANAGING THE MEADOW

Roffey Park Institute is a charitable trust that is recognized internationally for developing innovative learning approaches that enable individuals to achieve their full potential both at work and in their wider lives. Since the Institute was founded in 1946, the training courses and research have shifted with the economy to concentrate less on the factory as a workplace and more on the wellbeing of employees in businesses. The success of the Institute led its board to consider expansion, a project they knew would be a huge undertaking. 'We set a project budget and, as Roffey Park is a charitable non-profit-making organization without substantial free reserves, there was no possibility that this budget could be exceeded,' explains Val Hammond, who was chief executive at the time.

Working with a professional project manager, the Institute appointed architects who worked closely with the initial project team to understand the requirements and constraints. The brief was complex: redevelop the site of the Roffey Park Institute to replace existing residential accommodation with double bedrooms of four-star hotel standard, add a conference centre, provide additional dining and catering facilities, and incorporate suitable landscaping to link and blend the new and old buildings (some going back to the 1850s) into a harmonious whole.

'The architects' design exceeded everyone's expectations in terms of meeting the brief and also took into account the environmental issues and the need to keep running costs and ongoing maintenance as low as possible,' Hammond says.

But the project soon hit a major problem. Roffey Park is situated in the Sussex countryside in an area of outstanding natural beauty (AONB) – zones that are carefully regulated and subject to rigorous planning requirements. 'AONBs also tend to arouse strong emotional responses with the community, since, by definition, they are beautiful and rare places,' Hammond continues. 'In this case, the pre-planning application consultation with local residents took considerably longer than foreseen.' In fact, the consultation process delayed the original opening of the Meadow complex by around two years.

'We have been supporting the local community, as well as delegates from around the world, for decades, so it was important to us to do the consultation and application properly,' Hammond says.

Once planning permission was achieved, a permanent project team was appointed and everyone was confident of their ability to build within the budget figure. The team produced a new schedule for the project and agreed the dates. However, the Meadow project hit another problem as soon as the detailed building drawings were available, as the construction company quickly formed the view that it could not be built for the agreed budget. 'Prices had risen since the designs were developed, and they estimated the project, as planned, would now exceed the budget by nearly half as much again,' Hammond explains. 'We were not able to increase the available funds, so this was a much more serious problem than the time delay.'

The project team presented many options to reduce the cost, including losing the conference room and cutting the number of bedrooms, but the Institute was utterly committed to the flexibility the original design offered, and the building in its entirety had become integral to the business plan. The Institute, renowned for its innovative approach to management research, applied some of those skills to come up with a new way of bringing the project back under budget.

Hammond, now chair of the Roffey Park Institute, continues: 'Diligent work and painstaking negotiation on the part of all involved brought the project back into line. No elements were cut entirely, but some were reduced in scale. Elsewhere, detailing was streamlined to simplify the construction, but distinctive elements such as the egg-shaped pods to be used as syndicate rooms, the environmentally-sound construction and living roof over the conference centre were retained.'

In a project where there was no margin for error and very little in the way of a contingency budget, Hammond knew that keeping the budget on track for the duration was going to be a challenge. 'The principal actors – the Institute as the employer, the architect, the quantity surveyor, the construction company, the site and services engineers, as well as the project manager – accepted joint responsibility for the budget,' she says. 'We agreed that if any element went off track for any reason, there would be a collective effort to find a saving elsewhere, without compromising the quality of the building.' It was an approach that proved to be very successful. 'Sometimes it involved one discipline finding ways of offering more cost-effective solutions in their area to allow another to meet a necessary overspend that could not have been anticipated. On occasions, it meant the Institute had to compromise too.'

The project team worked extremely closely together with a high degree of trust, and there was architectural support on site continuously to deal with design queries and develop cost-effective solutions. Working in this way, the whole team brought the new buildings to completion at the agreed budget and on time. The Meadow opened in 2003 and was awarded a commendation by the Civic Trust in 2005. 'Even more important is that the building fulfils its purpose, the different elements work effectively and the facilities are always in demand,' Hammond says. 'Maintaining a clear focus on the primary objectives – a quality building that fulfils its purpose – and respect for each and every member of the team helped enormously in creating a sense of shared ownership and pride in our achievement. This, in turn, enabled us to achieve the impossible: to bring a building project to completion on budget.'

Budgets are, by their nature, collaborative. One person is not the root of spending all the money, even if one person approves all the expenditure. Different team members have their parts to play in recommending suppliers, keeping you informed about what needs to be purchased and so on. Encouraging everyone to work together to keep costs in check and to report accurately will generate a sense of collective responsibility for the budget within the team.

HINT

Present the overall budget to the team at the outset of the project. At this point, a high-level overview will do, concentrating on the overall amount to be spent and how that will be broken down. Explain the budget processes that affect the way the team works, for example that you expect them to use a certain cost centre code for their timesheets.

Sharing the project budget with the team has several positive effects. It:

- gives team members a sense of the scale of the work;
- helps them to feel part of the entire project instead of just their tasks;
- educates them as to the costs of their work in relation to the project;
- encourages them to report accurately as they see the impact their data has on the overall budget;
- highlights to the team the speed with which money is being used on the project;
- raises awareness of the costs related to each task;
- promotes a sense of individual responsibility (many team members will not have been entrusted with this type of information before);
- promotes a sense of collective responsibility (team members can see the interdependencies on other departments at a budgetary level);
- presents an opportunity for the team to challenge the budget or to add or revise estimates;
- establishes you as a project manager willing to share information, setting a good example for your team.

The budget may be fixed, but there might be some degree of flexibility about how it is spent. Having collective awareness and responsibility for the budget, even if you keep the authority for it, can prove a useful strategy when the finances are relatively flexible. Team members can really see how overspends in their area affect other departments.

Report the budget status at each team meeting: what is on track, which tasks risk going overspent and which look underspent? Present a recommendation for reallocating the budget and negotiate with the team until you reach a decision that works for everyone. Can a task be done by a less experienced (and therefore cheaper) resource? Can an expensive piece of equipment be replaced by something cheaper but of the same quality?

Reorganizing the budget within the team is one approach, but as soon as the solution to your budget concerns is to change the quality, alter the way in which the project will deliver the required objectives or change the timeframe you must get the sponsor's approval. Anything that fundamentally changes your time, scope, budget and quality 'contract' with the sponsor must be ratified by them and, if necessary, the steering group. J. Nevan Wright puts it like this:

> **If the sponsor determines that it is essential for the project to be completed by a set date and there is no flexibility or slack in the project, then extra resource and extra cost might have to be accepted. It is the project manager's responsibility to find alternative methods and courses of action in an endeavour to keep the extra cost to a minimum.**[12]

Chapter 13 looks at handling changes in more detail.

WARNING

Not all teams will respond well to this kind of openness. There is a risk that some departments might want to spend as little as possible in order to gain political 'points' and, in doing so, cut corners or report inaccurately, giving you a false impression of what has been completed. On the other hand, some team members might want to appear important and report that their workstream has overspent due to fixing a difficult problem or working extra hours. It is up to you as the project manager to gauge the reaction of your team and decide how much openness is appropriate.

When you are reporting your budget to the team, remember that in general people don't keep track of what they have done or how much time they have spent. How many times have you got to the till in the supermarket and not realized how much you have spent until the assistant asks for the cash? Your team may not realize how much of the available money they have already used up, and sharing the budget with them through regular team meetings is a useful way of educating them about project finances.

GOLDEN RULES

Although it might feel natural to keep the figures to yourself, sharing the project's financial status with the team on a regular basis can aid discussion and help facilitate collective responsibility for bringing the project in on budget.

7 Agree who holds signing authority

Signing authority is the ability to say yes to expenditure. A project budget describes and tracks all the things that need to be purchased, but someone actually has to sign the invoice, raise a purchase order and be accountable for the money that leaves the company. It could be you as the project manager or someone else; either way, when you buy something, you need to know whose desk the paperwork should land on.

a

TAKING RISKS

Marie-Hélène Dupleix, a project manager from Brittany, was working on launching an online employee-satisfaction survey for 6,000 staff across five office buildings when she hit a problem: 'The budget was already approved, but I needed someone to raise a purchase order for a new server. The technical team said we needed it quickly to stick to the schedule, so I approached my sponsor. He said it wasn't his role.' Dupleix spoke to all her key stakeholders and the finance department manager, but no one would take ownership of the budget. 'Asking around took me four days, and at the end of that we risked being behind schedule and I had achieved pretty much nothing except a headache,' she says. Dupleix took a risk and used a template she found on her word-processing package to produce what could pass as a purchase order. She added the project's budget code and sent it off to the suppliers. The server duly arrived.

'It took three weeks for the server to arrive, which was fine,' she says. 'During that time there were other day-to-day concerns and I really didn't give it a second thought.' But then the invoice came and the suppliers needed paying. Dupleix approached her sponsor with the paperwork and asked him to sign it. 'He asked me if we were within budget and if this was planned expenditure. I said yes, so he signed it,' she says. 'I sent the forms to the finance department and as I didn't hear back from the supplier I assume they were paid.'

Dupleix acknowledges this was a bit of a risk for her. The next time the project needed to spend money, she approached her sponsor again. 'He asked me what I had done last time, as he didn't remember authorizing any expenditure,' she explains. 'I told him that I had raised the purchase order as we needed to move quickly. He said I could carry on doing that and signing for project expenditure as long as we were within budget, so I did.' Dupleix feels she was lucky that it turned out this way: 'But sometimes you have to take risks to get results and this one really paid off,' she says. 'I got taken more seriously by my peers and the finance department, and I didn't have any more delays or headaches.'

The process for approving expenditure differs from company to company. A little research at the beginning of the project will let you establish the process you need to follow. There may be multiple steps in the chain, multiple signatures to get or committees to attend. Someone in the process may be very difficult to track down. The more process steps and the busier the people in the process, the harder it will be to purchase anything. The benefit of working this out at the beginning of the project is that you can build adequate lead time into your plan. Knowing that it will take three weeks to get the three signatories in a room together with the right paperwork means you can make sure you start early enough to avoid holding up the rest of the work.

'The ideal situation is for the project manager to have control over the budget,' writes Kim Heldman in *Project Management JumpStart*. 'This doesn't mean that you should have unlimited signing authority but you should be able to sign for normal supplies, contractor invoices, and so on.'[13] Heldman, the chief information officer for the Colorado Department of Natural Resources, advises project managers to agree with the sponsor and finance department a certain level of signing authority for themselves. This avoids having to go through the purchasing process for small amounts of expenditure. It is good advice and will certainly lessen the day-to-day financial headaches.

Giving project managers any degree of signing authority will be a no-no in some companies. Heldman believes that 'project managers who have no control over the budget should not be held responsible for budget mishaps.'[14] Your sponsor may not see it that way, however, and it would be unprofessional to ignore the budget completely just because you don't have the appropriate authority to sign off invoices. You can still track expenditure, log the requests for purchases and flag any concerns to your project sponsor, along with your recommendations for how they could manage the relationship with the budget holder.

GOLDEN RULES

Ask for the authority to make purchases against the project budget. If this is not granted, establish the process for gaining approval for expenditure.

8 Watch each budget line

Monitoring your project budget needs to be done on two levels: macro and micro. At a macro level, you are monitoring how much money has been spent and is available overall. At a micro level, your budget will be broken into different 'lines' – types of spending – for example, training or equipment purchase. Monitoring at this level relates to how much each line has spent and has left available.

At a macro level your project could seem very healthy, but if you look at the micro detail you might find one line using up much more than anticipated, putting the project at risk of not having enough money to go round.

a

THE BLACK-HOLE BUDGET

IT projects can have complicated organizational structures, as Emily Stevenson, a project manager in the financial services industry, found out. She was working on the launch of a new software application with a team of 11 spread over four sites in two countries. 'We were using an outsourced IT company,' she says. 'I was only supposed to speak to our dedicated project manager there.' The IT company had outsourced the infrastructure and hardware purchasing to another contractor. Stevenson did not have much contact with the external infrastructure team and received weekly status reports and cost updates from the IT project manager. All seemed to be progressing well until she realized that at the current rate of spending, the IT part of the project would soon be sucking up money from other lines in her budget. When she queried the costs, the infrastructure team's reports started to dry up. 'I was in the dark about the costs they were incurring,' Stevenson adds. The external infrastructure supplier invoiced the IT company monthly for all the work they undertook. Then the IT team passed the invoices on to Stevenson's company with their own fees included, so it was hard to break down the financial reports to get any accurate data. This was compounded by the fact that the IT team and the infrastructure supplier were the preferred partners for Stevenson's company and so worked for them on multiple projects at any one time. When Stevenson tried to find out exactly how much her project had been billed, she found that the IT company invoiced her company one lump sum for all the projects per month.

Stevenson ended up circumnavigating the IT project manager by working directly with the finance team to keep track of costs. By the time this was organized, the infrastructure team had already gone over the limit they had been approved to spend.

'The suppliers had spent money that we had never agreed to pay, just by racking up more resource costs,' Stevenson explains. 'Fortunately we had a watertight contract and they had to complete the work at no extra charge.

Because I had gone to the experts in finance, I found out my project didn't need to bear the charges for any overspend and we were able to stop paying,' she says. 'If I hadn't been paying attention to their charges and found a way through the system, the whole project would have been overspent just because they could charge whatever they liked and, for the most part, no one noticed.' After Stevenson's experience, controls on both outsourcing companies were tightened and a formal reporting structure implemented so project managers could accurately track the costs for their projects with more confidence.

At a micro level, it is more likely that costs relating to people will go over budget. Hardware purchases or one-off spends are normally fixed costs for which you will have accurate estimates. People costs are harder to estimate, and the human element means the staff working on the project might not know when they have reached the agreed limit.

HINT

Every hour spent working on the project is an hour out of your budget, so even working out your financial position changes it. Do financial reporting and tracking on a regular basis and report as accurately as possible. However, understand that the figures you provide in your reports are not going to accurately reflect the budget position as of the moment you press 'Send' to email the document to your sponsor. Explain the constraints of real-time reporting to your sponsor too. Unless you have a very advanced real-time tracking system within your company, they are going to be within the 60 per cent of decision makers who receive out-of-date management information.[15] They should be able to work with the data though: providing reports with data two months out of date is unacceptable. Your figures should never be more than a week old. As most time-tracking mechanisms work on weekly timesheets, this should be easily achievable.

It is also essential to keep an eye on what your external suppliers are spending, especially if you have a fixed budget. If you are expecting them to work within approved levels of cost, make sure they are aware of the amount and are prepared to stick within those limits. Agree in advance how they will report to you so you can monitor their spending: request copies of their timesheets if necessary. It will add an amount of extra work for you, but careful monitoring will make sure you are aware if things start to go awry. This will give you time to work with the supplier to bring the spending back in check, or to request additional funding if necessary, before the approved limits are breached. In a large project, you could reach those limits quickly if the supplier's estimates are inaccurate or based on a different understanding of the project scope. If your budget is more flexible or your project smaller, still consider the advantages of close cost monitoring. Keeping an eye on your suppliers' spending will ensure that they do not charge you extra simply because they rely on you and your finance team not to realize when they have charged more than you agreed at the outset.

Svetlana Cicmil sets out her definition of a project in *TQM Magazine* and says: 'Effective management processes of planning, monitoring and control are required to translate the idea of change into tangible deliverables.'[16] Budget monitoring and control processes will help you reach the end of the project in a healthy financial position and probably having suffered from fewer headaches along the way.

GOLDEN RULES

Monitor your budget at a micro level to:

- be sure that each budget line stays within the amount planned;
- give yourself time to correct anything out of line if there looks like there will be an overspend.

9 Arrange for a peer review

DEFINITION

A peer review is an informal audit that looks at the project so far through a pair of external eyes. Peer reviews are a useful tool to check the project is on track and give you confidence that you are doing the best possible job. They can be run by an external company, the internal audit or quality function of your company, or the project management department, with project managers carrying out peer reviews on each others' projects.

If you are offered the chance of a peer review, take it. If not, think about organizing one for yourself at key points during the project: insights from an unbiased evaluation really will be worth it.

CONSTRUCTING A SOLID PROJECT

Multiplex, an Australian property company involved with the building of the Wembley National Stadium in London, carried out a detailed review of the costs of the project in February 2005. Since that review, the company has put a number of new measures in place to mitigate risks on their portfolio for any contract exceeding A$100 million. One of these measures is a regular internal review.

The company carried out a peer review of the stadium project, which highlighted that the productivity levels previously assumed were not actually being achieved. The review also pointed out potential financial shortfalls, as the predicted project costs were more than planned. The internal review procedure concluded that the project would make a loss, leading senior management to implement strategies to tackle the project's financial challenges.[17]

A peer review is:

- an unbiased, friendly estimation of the project activity;
- supposed to point out areas where you could improve;
- going to offer suggestions to make things run more smoothly.

A peer review isn't:

- meant to identify mistakes;
- going to parcel out blame;
- intended to reduce your team to a quivering wreck.

Continuous improvement is part of the project manager's role, so where is the value in a peer review? Dan Bradbary and David Garrett answer this question in their book *Herding Chickens: Innovative Techniques for Project Management*:

> **After all, at the start of your project you established a set of metrics and other milestones to measure your progress and judge your own attitudes, no? True, but those are metrics that you've set, and set in your favour – otherwise put, they are biased towards you to begin with. [A peer review] brings a truly outside perspective to your project; it rids you of internal bias completely.** [18]

A peer review will normally look at the project holistically, but if you have specific concerns or doubts, ask the reviewer to pay particular attention to those areas. This is especially useful for budgets, where a second opinion will either reinforce your methods or identify ways in which you can improve. A review of your project budget can also give your sponsor confidence that the money is being handled appropriately while at the same time pointing out future shortfalls. 'If, for example, you've been pushing for a third network printer for two months but your pleas have fallen on deaf ears,' Bradbary and Garrett continue, 'a recommendation from an outside party may be just the ammo you need to get heard.'[19]

HINT

Expect to hand over copies of your monthly reports, your budget spreadsheet, plans and other project documentation during a peer review. Free up some time so you can talk to the reviewer and make your team available too if necessary.

The reviewer should produce a final report including their forecasts for the total expenditure and predicted finish date. These will either support yours or give you an alternative to consider. Any alternative view of your project forecasts should be backed up with some well thought-out reasons that will make it easier for you to decide whether you want to make changes to your plan based on the reviewer's report.

HELP! I'VE BEEN ASKED TO BE A PEER REVIEWER

If you are asked to be a peer reviewer, be flattered. Someone thinks you are a good enough project manager to pass judgement about another project, objective enough to present your findings in a clear and blame-free way and detailed enough to uncover things that they might have missed themselves.

Being a reviewer is not a daunting task. Find out whether there are standard templates for carrying out peer reviews already in use by other project managers. Even if your search doesn't turn up anything official, speak to someone who has done it before. Think about what you would want to know if it was your project: is the budget on track, is the schedule realistic, how are risks and issues managed? You do not need to be an expert in the technical subject matter of the project to carry out a successful peer review, but there is an obligation to give your feedback in a constructive way. Focus on constructive criticism and suggestions for improvement rather than highlighting errors the team can now do nothing about.

Having a final report from a peer reviewer is a great start. To make the whole review exercise worthwhile though, you really need to put those recommendations into practice. Read through the recommendations carefully and work out how and when they can be implemented: obviously, the sooner the better. Ask the reviewer's advice if you cannot see how to turn the recommendations into reality. They will have suggestions for a successful implementation that might help you see the problems in a different light.

GOLDEN RULES

Schedule peer reviews at regular intervals during your project, and be sure to act on the reviewers' recommendations.

10 Manage the model

Your boss presents you with a complex spreadsheet or a number on the back of a postage stamp and says: 'That's the way we always plan budgets.' It's a common scenario: the company has been doing things this way for years, and no one really remembers why. It's just the way things are. But what if the corporate way is not the right way for you? The principles of budget setting and monitoring discussed in this section remain the same, whatever methodology is proposed. You will be surprised at how much flexibility you can find in budget processes, which is just as well, as you will no doubt be working with disparate teams, using different calculation models and with different levels of understanding about how budgets work.

Having the confidence to take a slightly different approach from the corporate norm may mean the difference between actually understanding what you are doing and letting money slip through your fingers.

a

IT'S DOWN TO YOU

Ketan Samani, a senior project manager in the financial services sector, has handled many budgets. 'In my opinion, good budget control and monitoring spend during the project lifecycle are not down to a model or methodology, but a good project manager,' he says. 'Their personal qualities and ability to negotiate are important.'

Samani is an ecommerce specialist with a background in sales and marketing. His experience managing projects has taught him the value of negotiating budget constraints alongside other aspects of the project. 'At best, again in my opinion, most projects are started without a comprehensive study of the project's purpose, and therefore the value they can deliver is always limited,' he explains. 'This can happen consciously or unconsciously, depending upon the sophistication of the team managing it.'

Samani is working on implementing a project management framework and governance model in his company that advocates the use of monthly budget reports based on a standard template. However, he is clear about how to get the most value from tools like these. 'One can provide on- and off-line project management models and tools for budget management, but it is down to the operator to use them efficiently,' he says.

Many 'flavours' of budgeting are used within organizations. Some will require a full business case before any work is done, including net present value calculations, payback periods, internal rates of return and all manner of other accounting principles that another book can explain better than this one. Some systems are hideously complicated and come with a health warning: you risk spending your entire time making sure figures are entered into the right boxes so that ten other spreadsheets can update automatically. That's fine if you enjoy that sort of thing, but don't let your passion for an accounting package undermine your ability to do the rest of your job. Other organizations will expect quarterly forecasts. Others will be surprised if you even ask the question about what system they normally use.

HINT

Whatever the methodology advocated by your company for setting and managing project budgets, make sure you understand it. And if you don't, at least understand the basic principles as a minimum and devise your own way of recording the relevant information that satisfies your needs as the project manager and the needs of the finance team or whoever has authority over the budget. Do the minimum for the 'official' methodology and complete your monitoring and tracking with your own method.

You may need to find a way to complete or modify your company's methodology for another reason: it might not be sufficient for dealing with the type of project you are working on. Project budgets often come from various different places. Head office might be funding half of it, with the rest paid for by the local office. Your budget might be in a mix of currencies or spread across two or more financial years. In research done with a group of project managers, operational managers and team members, the most frequently cited limitation of the methodologies and tools they were expected to use was their inadequacy for handling complex projects.[20]

TIPS FOR GETTING TO GRIPS WITH BUDGET METHODOLOGIES

- Don't panic about the maths: not all project managers are born mathematicians. If you don't understand it, ask for help. Budgets at their most basic don't rely on complex maths: you have a sum of money and you spend it over time, just like a household budget.
- Question why certain elements are required. If the answer is unsatisfactory, you can get away without doing that particular report or sum. If the answer is satisfactory, you will have a better understanding of what the finance team really needs.
- Remain true to the principles of the methodology, but feel free to adapt it to suit your needs. There is no need to do net present value calculations for a project that will be complete in three months.
- Team up with someone who has used the methodology for a project before and find out what worked well and where you will have problems. They can offer useful insights into where the model breaks down.

The key elements to be aware of are:

- Do you know how much money you have overall?
- Do you know how much has already been spent?
- Do you know how much you plan to spend?
- Do you know how to address the balance if you plan to spend more than you have?
- Are your budget-setting and monitoring processes efficient?
- Can you explain all these points to your sponsor and answer their budget questions intelligently?

If you can answer yes to all those questions, then you have found a suitable way to manage your project budget.

GOLDEN RULES

Use the tools you are obliged to use in the most efficient way, even if it means bending the corporate rules.

11 Manage projects with no budget carefully

You can buy whole books about risk management, critical path analysis and soft skills, but there is very little written about managing project budgets other than the odd chapter in general project management books (like this one). That is symptomatic of the fact that the majority of small projects, especially those without IT involvement, are run without a specific budget, around the edges of a manager's day job. Any costs have to come out of the business-as-usual provision, which means there is no allocated project budget. So when your project has no budget, how can you keep a budget on track?

a

NOT COUNTING THE PENNIES

'When our branch manager told me I had to set up an online system to run the quarterly quality assessments for call centre staff with no budget, I admit I wasn't happy,' confesses Gordon Harvey, an IT project manager at a travel firm. 'Partly because I didn't think it would be possible and partly because I'd just finished running a project with a budget of £100,000, so I felt it was a bit beneath me.' The company's customer-facing staff undergo a rigorous assessment every three months to check they are providing a top-quality level of service to travel customers. This process took up a lot of time for team leaders, and the paperwork involved made everyone dread test week.

Harvey thought creatively about what could be an appropriate solution and drafted in a part-time team member from IT to help. 'We were a project team of two – me and one other – and we were both only working on this project two days a week,' Harvey explains. They needed to find something simple, quick and, more importantly, free. The obvious solution was to take advantage of the company intranet, but it had never been used in that kind of way before. Harvey was unconvinced that his IT colleague had the skills to build something complementary. 'By chance, I was talking to someone else in a different branch about a different project, and he happened to mention they were in the process of building a database to hold customer satisfaction scores,' Harvey says. He immediately saw a link between the customer survey data and the type of service questionnaire the call centre agents completed four times a year. The database was being created in conjunction with the marketing team who were managing communications to the customers. Harvey put a call in to the internal equivalent – the internal communications manager. 'She was interested and helpful in a reserved way,' he says, 'but didn't have any resources available to help practically. What we really needed was a comms professional who could spend some time with the marketing people from the other branch to find out how we could apply the same logic to the customer service questionnaire.' Harvey got on the phone to his sponsor and explained the difficulties they were having. Some negotiations took place at a senior level, and Harvey found himself with the comms person he wanted working on the project for a day a week. In return, he later found out, his sponsor had promised to move the resolution of an outstanding technical issue that was affecting the internal communications team higher up the priority list. Harvey got the specialized resource he needed, and internal communications benefited too: no need for money to change hands or resources to be formally purchased, and the project stayed within the day-job confines originally prescribed – only with a little extra help.

Projects with no specific financial amount attached to them are normally expected to be delivered using just the resources available as part of a day job. That basically means drawing on the people around you to do whatever it is that needs to be done. Having a project with no budget takes away some of the financial headaches but doesn't mean you are in for an easy ride. You will still have deadlines to meet and requirements to deliver.

WARNING

In some respects no-budget projects are harder to deliver, as you cannot throw money at a problem to make it go away; nor will your team have access to overtime payments if things start to slip.

If your sponsor hands you a project and then says 'There's no money available to do this,' count to 10 and try to avoid spitting 'You must be joking!' Ask them how they expect it to be achieved. A sponsor who is serious about a project will already have thought about what they consider to be a reasonable investment for a successful delivery. Take them through the questions below and start to work out where your boundaries are, particularly in relation to the people to whom you have access and the amount of time you and they can be expected to spend working on the project.

SPONSOR QUESTIONS FOR NO-BUDGET PROJECTS

- Am I full-time on this project?
- If not, what percentage of my time do you expect me to spend on this?
- Do I have any full-time resources?
- If not, what percentage of their time do you expect them to spend on this project?
- For any resources not under your control, has their manager agreed that they will be working on this project?
- Can any costs come out of the business-as-usual budget?

- To what limit?
- Who will authorize this?
- If the business-as-usual budget is not available, how do you want me to deal with unforeseen actual expenditure?
- At what point does the project become unfeasible?
- When does the resource investment become too much for your intended deliverables?

Once you have a clear idea of where your sponsor believes your boundaries are in terms of consumable resources, both business-as-usual budget expenditure and time, you can begin to work on the project within those constraints.

HINT

Make sure any assumptions or constraints are written into your project initiation document. For example:

- The business-as-usual sales budget will cover the cost of reprinting a new edition of our catalogue.
- The schedule has been produced assuming no overtime is available.
- All resources will be available as necessary.
- The system changes can be achieved using the maintenance budget.

WARNING

When you are not buying your resources formally, and they work for someone else, there is a risk that their own day job will take priority over the project. Despite good intentions, there will be times when staff shortages, increased workloads or other short-term crises drive your team back to their normal activity. If you can, schedule contingency time to keep your resource planning flexible. Always include an item in your risk register about the possibility of resources being pulled off the project. As a minimum, each time you review the register, it will prompt you to look at the current situation and see whether you need to take any action.

WARNING

If at any time the project looks like it will have to spend real, tangible money and you don't know where it will come from, raise this immediately with your sponsor as an urgent issue.

GOLDEN RULES

Even if no specific budget is available for the project, clarify with your sponsor what they consider to be a reasonable investment in terms of time and work within those documented constraints.

Section 2
Managing project scope

A man steers well who reaches the port for which he started.

Lucius Annaeus Seneca (c. 4 BC–AD 65), *On Benefits*

Scope forms part of the golden triangle of project management, along with resources and time. Figure II explains the triangle. Scope, resources and time are balanced equally within a project and form an equilateral triangle. If there is a change to any part of the triangle, another element has to change as well to keep the balance. For example, if scope increases, time has to increase in order to give the project team longer to deliver the proposed changes. If this is not possible, then more resources have to be found in order to do an increased amount of work in the same amount of time. As with all theoretical models, the practical application is perhaps more pragmatic. The golden triangle assumes that quality is a constant, but a pragmatic solution to achieving things faster and more cheaply is to deliver to lower standards of quality.

Scope is perhaps the most fluid of the three elements, as changes are an inevitable part of project management. The average project goes through four formal versions of scope and ends up achieving only 93 per cent of what it set out to deliver.[21]

This section looks at why scope management is so important and what project managers can do to avoid the traps of assumptions, implied organizational knowledge and scope creep.

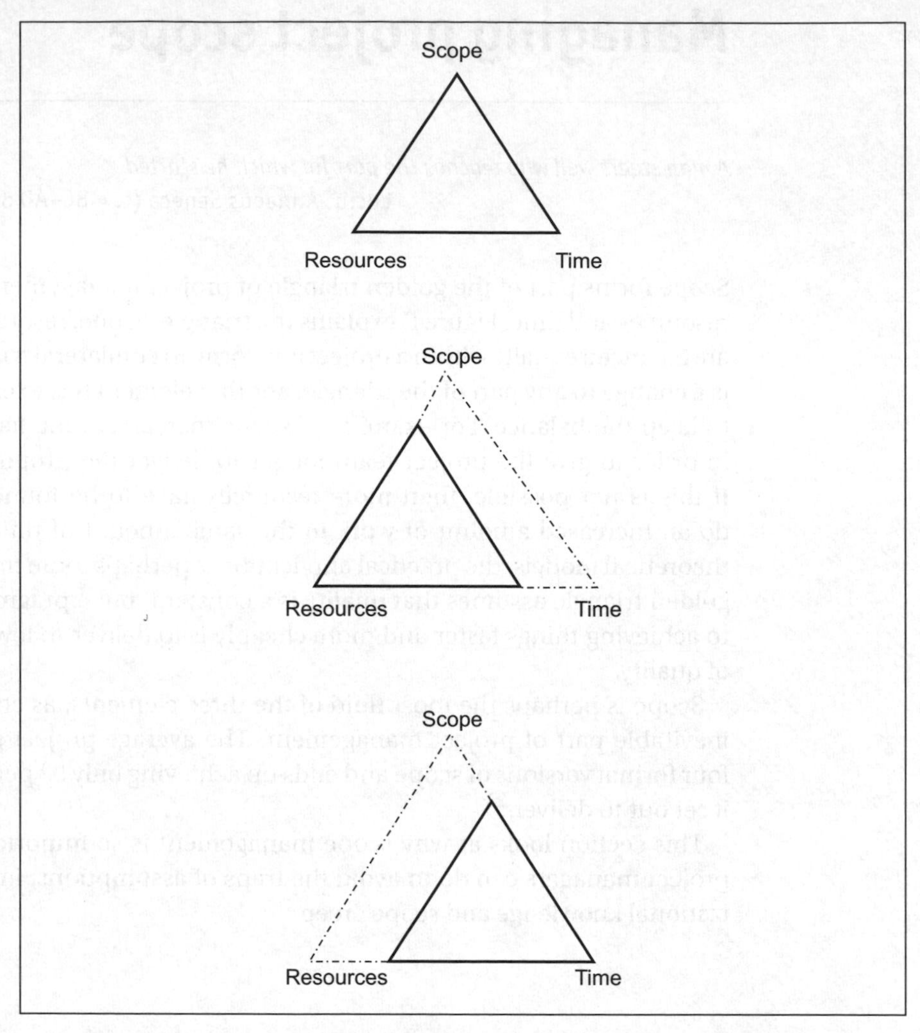

FIGURE II *The golden triangle of scope, resources and time management*

12 Keep it small

It is very tempting at the beginning of a project to include loads of things in scope and plan a big-bang deployment. However, this is rarely the best way to deliver change, especially in the IT arena. Where possible, packaging the change into smaller, interim deliverables supported by full testing is a safer and more robust approach to implementation.

a

FROM SMALL ACORNS . . .

When Paula Gardner set up her public relations (PR) business in 2002, she started small. In fact, she started marketing her etraining course in PR for small businesses even before it was written. This step-by-step approach has let her grow her business in small increments, learning from the experience as she went along.

Once a few clients had booked that first course, aware that it was part of a pilot, Gardner wrote the course material and launched www.doyourownpr.com properly. She used the pilot group to refine the course. 'After the course, I asked for feedback by email and about 70 per cent of participants replied,' she says. 'By leaving it open, I got a great range of answers. My clients are time-poor so I don't get a good result from structured questionnaires.'

The pilot and the feedback Gardner got afterwards allowed her to refine the ecourse in two ways. First, she honed the content, including things people had asked for and making sure all the terminology was explained clearly. She made sure she was delivering exactly what her clients wanted. 'I went right back to basics after reading the feedback,' she says. 'For example, I assumed people knew what a press release was, but in response to my clients' comments I made sure I included a clear description.'

Second, Gardner made changes to the way in which the course was rolled out. She put up the price to £99. All the pilot participants had to start the modular course at the same time, and Gardner emailed them instalments as a group. The next group of customers had to wait until the next course was due to begin before they could start. 'I invested about £2000 in my website and software to enable clients to enrol and start the ecourse on a date that suits them,' Gardner says. 'Many clients sign up on a Friday and want to start on the Monday, for example. Now they can.' This change to the deployment methodology made a huge difference to the flexibility and marketability of the course. Gardner supplements that by ringing every client when they enrol to help them make the best of the ecourse.

From a small initial rollout, Gardner has seen over 450 clients take the ecourse in PR. She has also used continuous feedback to launch other courses in networking, writing articles for the media and newsletters, with each launch drawing on the experience of the previous ecourses. The company has continued to grow, with Gardner adding coaching and consultancy to her portfolio.

The key to getting it right first time is keeping it small to begin with. When you plan the scope of your project, do not be tempted to include too many things. You can always have a phased approach, where your first project delivery (and so your first scope statement) covers a small portion of what you would ideally like to do, and then add subsequent phases to manage the rest of the rollout. There are very few projects that would not benefit from a pilot or proof-of-concept stage at the outset, so consider including one in your project scope. Rolling out a new system to a handful of users, asking them to test it rigorously and then making changes based on their feedback, is a lot easier than deploying the system to everyone and then managing the internal communications, bad publicity and general ill-feeling when you implement the fixes later. Your users will be asking themselves: 'Why would this latest update be any better?'

HINT

Piloting is also a way to contain costs. Identifying problems early gives you the chance to manage them when they are still small and can be corrected without significant cost.

THE SIX STEPS OF PILOTING

(i) Establish how a pilot could add value to the project by mitigating risk.
(ii) Get commitment from relevant stakeholders.
(iii) Define the evaluation criteria for the pilot: how will success be measured?
(iv) Produce a project plan and other project management deliverables.
(v) Deliver the pilot: run it for a predetermined length of time.
(vi) Evaluate the success based upon the criteria established in step (iii).[22]

Although being able to test functionality is normally the main reason for a phased deployment, keeping it small also allows you to learn from the implementation process. The system might be great, but could you tweak the training material? That branch opening was a success, but when the bigger branch opens could you get more local press coverage? The new product is selling well to a trial group, but how can you act on customer feedback regarding service? Take the lessons from customer feedback about the end product and the feelings of your users with regard to the deployment process to inform and improve the next phase of the rollout. Complaints are a gift: someone has given you the chance to tweak the product and get it right before you unleash it on your customers and to confirm that all the snags in the implementation plan are ironed out before you go for a huge launch. The benefit of starting small is that you have the opportunity to make things better for next time. Use it!

GOLDEN RULES

Putting a pilot or proof-of-concept stage in the scope of your project and/or planning a phased rollout will allow you to fully test both the deliverables and the implementation approach, so you do not lose the support of your project customers by delivering something that is not fit for the job.

13 Work out how to manage changes

Change is an inevitable part of projects. The whole point of a project is to change something. There are two types of change:

- changes that come about as a result of the project, i.e. the deliverables of your project;
- changes within the project, e.g. as a result of new requirements, upgraded technology or external pressures such as new laws or modifications to regulatory frameworks, or a sponsor who changes the priorities.

The impact of the first type of change is the subject of your project and has been carefully planned with a communications and implementation strategy. It is the second type of change with which we are concerned here: how to manage changes that affect your project objective in some way. Advance planning for how to handle these changes will make it easier when one comes along during the project.

a

KEEPING IT UNDER CONTROL

Process-driven industries rely on accurate monitoring and careful management of changes, as Leo Dijkhuizen knows well. He has spent 10 years in a chemical and pharmaceutical environment in the Netherlands, where strict change-control procedures are a necessary part of his role as a project manager specializing in complex accounting software implementations. Working on projects with budgets that stretch into millions of dollars and timeframes of several years, he has found it absolutely critical to manage changes in a controlled way.

Dijkhuizen explains the change-control process of a project to define and implement a global SAP system: 'Each change had to be justified with a business case,' he says. 'The steering committee reviewed and approved – or rejected – each business change.' The system was being rolled out across six European countries and the USA, with the aim of globalizing the company's supply chain and lowering inventory costs.

The project took three and a half years, and with a budget of $32 million it would have been easy for small changes to mount up over time and push the scope and budget way over what had been anticipated. 'The customer required a tight control over changes, so it was not difficult to implement a change-control process,' Dijkhuizen says. Having global, centralized change control helped to achieve a smooth rollout across the seven countries. The steering committee signed off each change and was sure it understood the implications.

Dijkhuizen's advice is to find someone in the project structure who can act as a sponsor for the change process. 'Make sure this person operates at the right level in the organization. If you cannot find someone who meets the description, ask the steering group what their policy is regarding changes and develop a process to manage each change around that.'

Change control or change management is the process of managing unplanned but desired influences on the project. It is important because any change will:

- need to be analysed for its impact on project objectives;
- need to be analysed for its impact on project scope;
- modify your existing plans;
- need to be recorded properly for a complete audit trail.

THE EIGHT STEPS OF A CHANGE-MANAGEMENT PROCESS

i. **Document the change request in the change log.** Ask the person who has suggested the change to be as specific as possible and put the change in writing. If they have any supporting materials (such as quotes or estimates for the work that needs to be done) that might help the analysis, ask for those too. The change log is like a risk or issue log and in its simplest form is a document where changes and activities are written down. See the appendix for an example.

ii. **Analyse the request briefly to establish ownership.** As a team, look at the change request to decide whether to take it forward. This gives you the chance to discuss the political motivations behind the change and to throw out anything that is blatantly impossible. The change request process can be used by stakeholders to bring things to the project team's attention or to raise issues, so you also have the opportunity to check whether it is a legitimate change or a query that is better handled via a different route.

iii. **Assign an owner who can analyse the change impact more fully.** Step (ii) will have flagged who is best suited to carry out the full impact analysis. Allocate an owner to the change – someone who can coordinate this activity.

iv. **Assess the priority of the change request.** Give the change request a priority. Is it critical, important or nice to have? This provides a sense of urgency for planning the impact analysis.

v. **Analyse and report back on the change impact.** The relevant people should carry out the impact analysis looking at:

   - what elements would need to change;
   - what the impact would be on the schedule/budget/resources/quality/business case (remember the contingency budget is not there to pay for new requirements!);
   - what effort would be required to make the change;
   - what the impact would be if the change is not made.

The owner should report back to the project team (larger projects may have a delegated change control team or forum) with a recommendation.

vi. **Decide the course of action: approve or reject the change request.** Take the decision, ratifying it with the relevant stakeholders and those affected by the change.

vii. **Record the outcome in the change log.** Complete the documentation for the change log. Note whether the change request was approved, the rationale behind it and the date the decision was communicated to the requestor. Include where the impact analysis can be found so if there is a query or similar change raised later you can find it again easily.

viii. **If approved, update all relevant documentation.** Cascade the relevant changes into all the project documents that need updating, such as the project initiation document, plan and budget. Issue a new version of those documents so everyone has the current view of the project scope and requirements.

Your company's process may be slightly different. Follow internal guidelines if there are any, but the activities will be largely the same as described here.

Change management can be unsettling. You start out on a well-defined project and suddenly everything is different and there is a stack of paperwork to complete for the audit trail. It is not all bad news though. 'Change requests are cause for celebration,' says Bill Duncan, director of standards for the American Society for the Advancement of Project Management. 'When your customers or clients ask for something new or different, it means that they are actually involved in the project. It means that they care about the project, that they are still interested in it, and that they are still planning on using the product of the project.'[23] Looked at like that, change requests are actually good news, and once you have been through the process for the first time, handling subsequent change requests becomes much easier.

GOLDEN RULES

Establish a change-management process following the steps outlined here to handle deviations from the original project scope, requirements, schedule or budget.

14 Include quality planning in scope

DEFINITION

A quality plan is a document that describes how quality is to be achieved during the project. Like any plan, it includes timescales, milestones and resources: namely, what quality activities are to take place and who will do them.

BANKING ON PROJECT MONITORING (PART 1)

Ian Duthie, a senior project manager at Lloyds TSB, was working on a human resources (HR) project with a vast scope, including payroll integration, web-enabled direct access for some HR processes to employees, web-accessed processes for line managers and enhanced management information. The project was part of a larger programme aiming to extend the model of using just one HR department for the entire group – no mean feat for a company operating across 27 countries, with subsidiary companies Cheltenham & Gloucester, Lloyds TSB Scotland, Scottish Widows and Goldfish. The project was scheduled to be implemented in a series of releases and involved a team of around 160 people.

On a project of this scale, monitoring the quality of the deliverables was going to be a large job. The quality plan set out the standards for the project and how and when quality deliverables such as project health checks and quality assurance reviews would be built in to the overall activities. 'This was a project that had been running under an external project manager,' Duthie explains. 'When it was brought in house, we applied our own internal controls to it, one of which was a quality plan.'

The team was fortunate in that it was able to reuse parts of a quality plan from another project. 'We based the plan on something we'd done for another part of the organization on the same project,' Duthie says. To get the quality plan for the £50-million project recognized as part of the 'official' documentation, it was essential to have it signed off and approved by the teams involved. 'The overall plan was signed off by the programme manager, while the various deliverables were signed off at the appropriate level within the business and IT departments,' Duthie says. 'Some of the approvals took place by way of a "desk check", where the relevant people read through the document. Other approvals were more formal and were reviewed in meetings.'

Having a plan is only half the work: Duthie knew that getting the team to stick to it was going to be a challenge. 'We did stick to it!' he says. 'The fact that focus was kept on the quality plan was probably due to the nature of the people who were tasked with ensuring that the various deliverables were met.' The project team linked the quality plan to the project deliverable plan and made sure that major points for quality signoff were documented in the main project plan. 'As deliverables were due at various stages throughout the project, there was always a focus on the deliverable and milestone plan, and the quality deliverables just kept on coming up!' he adds.

The quality plan should include:

- any standards that must be adhered to;
- quality control and quality assurance methods;
- responsibilities: who will carry out the activities;
- quality tools (if you are going to use any);
- a reference to the change-control process that the project will follow.

The quality plan will also reference the project acceptance criteria and may include the configuration management procedures too.

DEFINITION

Quality control is the day-to-day activity of making sure the work delivered is up to scratch. It is done by the project team.

DEFINITION

Quality assurance is normally carried out by someone outside the project. It is an independent check.

There are two main parts to a quality plan: the definition and standards, and the schedule.

The first part of your quality plan sets out your definition of quality. Quality is a very vague concept. Svetlana Cicmil asks a good question in *TQM Magazine*, referring to the golden triangle we saw in the introduction to this section: 'If meeting time, budget and specification requirements within a given scope of project work are always a matter of trade-off among these variables . . . how should quality in a project situation be defined?'[24]

Each project and organization will have a different definition of what quality means. You will need to decide what it means in relation to your project. This can be a difficult task, so start by thinking: 'How will we know that what we deliver and the way in which we deliver it will be good enough?' Dig out some corporate documentation or ask around as a starting point using this checklist of questions:

- What project-management standards or methodologies are we expected to follow?
- What IT coding standards exist in the organization? For example, how will code be checked and tested before being implemented?

- Are there any industry standards to follow?
- Are there health and safety considerations to meet?
- What other legal requirements or generally accepted norms do we have to deliver to?
- What standard does the customer expect from a final product?
- How will these standards be measured?

Documenting these points forms part of your quality plan: setting out the standards to which you choose to work. If there aren't any, get your thinking cap on and come up with a framework that your team can agree to work within.

The second major part of your quality plan is a schedule of when quality activities will happen. Some will be integral and ongoing, for example proofreading a document before sending it out for comment. There is no need to record these. For each major project deliverable, specify:

- what will be tested;
- when it will be tested;
- who will do the testing;
- any tools that will be used;
- how the results will be recorded.

Also in this section include any planned dates for quality assurance reviews and who will be responsible for coordinating them.

The change-control procedure either will be documented fully in the quality plan or readers should be pointed towards another description of how changes will be managed. For more on change control, flick back a few pages to Chapter 13.

Delivering a quality product at the end of your project, whatever that may be, will go a long way to securing your reputation as a competent project manager. However, you may want to keep your quality activities low-key. Lynn Crawford discovered that paying attention to quality is a characteristic top managers associate with low performers. Managers in her 2005 study saw top performers as people with skills in many other project management areas, but being good at quality management actually decreased the likelihood of being seen as a great project manager.[25]

GOLDEN RULES

Quality is something that must be worked at throughout the life of the project, and having a quality plan will define how that can be achieved.

15 Work out how to track benefits

Projects normally aim to deliver some kind of beneficial change. But how do you know whether your project has been beneficial?

DEFINITION

Success criteria are the standards by which the project will be judged to have been successful in the eyes of the stakeholders. It is these criteria that must be tracked in order to answer the question of whether your project has delivered any benefits.

BANKING ON PROJECT MONITORING (PART 2)

The HR project with which Ian Duthie was involved was strategically important to Lloyds TSB. As one of the objectives was to make the bank's HR processes more efficient and effective for employees, it was vital to make sure the benefits were measured to find out whether the improvements were making a difference. As a result of the project, line managers and employees would be given direct access to HR information relating to their employment with the bank – a huge challenge for a bank with over 69,000 staff in 27 countries.

'We had a benefits management tracking system in place and devoted one person full-time to managing it after the main delivery stages of the programme,' Duthie explains. As well as bringing all the HR teams together, the programme aimed to develop the IT infrastructure underpinning the department's strategy. There were plenty of benefits that needed to be monitored to show that the project had been successful and pick up any areas where there were opportunities for further improvements. 'We wanted to ensure that we tracked benefits accurately,' he adds.

Duthie believes that benefit tracking is a part of the project team's responsibility. 'In this project, though, the ultimate accountability for the delivery of the benefits remained with the sponsor,' he says. 'The company will make sure the project benefits continue to deliver by writing the achievement of benefits into the targets for the year. So the achievement or otherwise of the benefits will therefore be a prominent feature of the executives' reward package.'

He has some advice for project managers, based on his 20 years of experience in financial services in the UK and overseas, most of it spent in a project environment. 'Make the decision of whether you will devise, adapt or adopt a system for benefit tracking and then use it!' he says. 'Think very carefully about the metrics at the outset of the project, and be sure of what you are measuring. Consider why you are measuring these metrics: will they actually help you prove whether the project has met the objectives set out at the beginning?'

Duthie adds: 'How are you going to measure the metrics you have identified? Where will the information come from? Is it readily available?' If the information you need is not available at the moment, then Duthie's advice is to think carefully about how much effort it will take to gather the necessary data. Benefit tracking is a balance between having enough information to ensure the project is, and continues to be, successful and creating an industry just to measure the benefits.

'Who will collate the benefit measures?' asks Duthie. 'And when is the measurement data going to be available? If the different metrics are not available at the same point in time and cover the same period – weekly or monthly – think about whether discrepancies in the data collection will cause problems or disputes. The last thing your project needs is people looking to exploit timing differences in reporting cycles to justify why certain targets were not met.'

Benefits are tricky, as it is not normally the project manager who has to track them. After all, you deliver the project and move on to something else. It is the operational business-as-usual team that has to live with the change and use it to generate benefits for the organization on an ongoing basis. Projects that are part of a wider programme of activity may be able to leave benefits tracking up to the programme manager. However, on large projects with several delivery phases, benefits tracking will fall to you, as you monitor the success of the previous stages in order to better deliver the later stages.

HINT

Even if it will not be you tracking the project's success, it is a courtesy to the business-as-usual team to put in place a mechanism by which benefits can be tracked – if only so that career-influencing managers will be able to see the successes you helped them deliver six months later.

Terry Cooke-Davies carried out research at more than 70 large organizations to come up with the 12 factors that are critical to a project's success. On the list is 'the existence of an effective benefits delivery and management process that involves the mutual co-operation of project management and line management functions.'[26] The way that process looks really depends on the project you are managing. Terry White sets out a straightforward two-step approach in his book *What Business Really Wants from IT*:[27]

(i) Identify what you want to measure. If the criteria are complex, such as employee morale, break them down into tangible, measurable chunks.
(ii) Establish the current baseline so you can track the improvement.

The first step in this approach is the identification phase. While you are brainstorming how you will know whether your project has been a success, you'll probably come up with success criteria related to the management of the project, which you can refer to in project audits or the post-project review. They will be useful to help focus your mind on the business of project management and relate to doing the project right, completing what you set out to achieve within the defined parameters. However, these are not sufficient alone. Your identification phase should also identify deliverable-based success criteria that are linked strongly to the business case and the rationale behind doing the project. There is no point in having success criteria such as 'ensure every project manager takes a full hour break at lunchtime' (unless you are running a project on improving the work/life balance and feel like being dictatorial). Your sponsor and business team have engaged you to do the project to deliver something that will benefit them, and that is what should be tracked. These two types of success criteria are summarized in Table 15.1.

TABLE 15.1 *Types of success criteria*

Criterion	Example
Project: things related to the professional job of running the project	Produce and gain signoff for project initiation document
Deliverables: things delivered as a result of the project	Distribute 3,000 educational leaflets to schools in our county

Success criteria can be measured in two ways:

- Discrete: yes/no. We did or did not do something. Examples: project delivered on time, company gained XYZ accreditation, new branch opened.
- Continuous: measurable on a scale. We did something to a certain extent, within a target range. Examples: improve customer satisfaction scores to between 75 and 100 per cent, increase revenue by 8–10 per cent, rebrand 20 offices between October and December.

Project-management-related success criteria do not need to be tracked over time, and so you do not need to generate a baseline of current performance. Once the project is over, you should be able to say with certainty whether, and to what extent, you met the criteria. The true business benefits, on the other hand, may last for a lot longer. Even a one-off project such as changing all the office lightbulbs to energy-efficient bulbs has durable benefits. The success criteria could be 'maintain electricity savings at 40 per cent of previous expenditure'. Tracking the benefits will make certain that the business-as-usual team will be aware when the costs start to increase again – and be able to find out who replaced a dead bulb with a non-efficient one.

A baseline of current performance should be taken before, or in the early stages of, the project. It should record the current performance against the success criteria before the project is delivered. This baseline gives the business-as-usual team something to compare against. It is great knowing that you are now calling back customers within 30 minutes, but if you do not know what the situation was before the project was implemented, it is impossible to judge whether things are better now. The baseline allows clear identification of performance differences in the post-project world. Be sure to use the same calculations and tracking method to establish the baseline as you plan to do for the ongoing measurement, otherwise you risk comparing apples with pears.

Continuous success criteria always include the possibility of being translated into discrete targets. If customer satisfaction was 82 per cent in March, and the target was 75 per cent, then you reached the target. If customer satisfaction was 74 per cent, then you didn't. Monitoring benefits on a continuous scale is always better, as it allows you to track changes over a period of time. If the customer satisfaction target was reached in April, then that's fantastic. But you cannot tell from a yes/no measurement whether it was better or worse than the 82 per cent reached in March.

We often talk about baselines in relation to plans, but baseline data – establishing the situation at any given point – can also be used across other parts of the project management lifecycle. For example, if your project is supposed to improve something, take a snapshot of the current performance before the project is implemented and then another afterwards. The first snapshot is your baseline; the second allows you to see how much progress has been made.

a

ANECDOTE

The Middle East Partnership Initiative (MEPI) was set up in 2002 to promote democracy in the Middle East and north Africa by providing assistance for political, economic and educational reform and women's empowerment. In 2005, their project-management approach was audited: the results showed clearly that MEPI had difficulty working out how successful its initiatives were, as there was little baseline information. The lack of baseline data meant that measuring performance was difficult. The auditors found that of 25 projects, only three project-initiation documents mentioned the need to establish a baseline and none of the projects had reported any baseline data at all. Unsurprisingly, the auditors concluded that 'without the ability to evaluate its projects' performance with certainty, and lacking access to complete information, MEPI's capacity to meet its strategic goals of producing tangible results and making results-based decisions is limited'.[28]

+

HINT

Identification of success criteria and the ensuing measurement of benefits should not be done in a vacuum. Dieter Fink's empirical research into benefits management led him to conclude that success criteria are 'driven by management and rely on open communications, transparency and recognition of diversity. Furthermore, only the involvement of all stakeholders will ensure that sufficient expertise and support exist to identify successfully and realize the benefits.'[29] Involving your stakeholders in the identification process will increase everyone's confidence that the chosen success criteria are actually the important ones. Involving stakeholders in the measurement of benefits will help to guarantee their commitment to the project outcomes.

GOLDEN RULES

Identify the project's success criteria early on and then track against a baseline performance target to measure benefits.

16 Eliminate ambiguity

On an IT project, you will have probably lived and worked with the details for weeks, if not longer, before handing over the requirements document to the team that will eventually build the system. It is easy to assume that as you know exactly what you mean, they will too – but this is a dangerous assumption to make.

a

ADDRESSING THE PROBLEM

Sarah, a project manager in the financial services sector, collated the requirements from her business users for a new IT system. They included the need for staff to be able to record clients' address changes. Clients wanted their post to go to their old address until the day they planned to move house, and then from that date to their new address. The customer-facing team decided it would be easier to tell the system to store the new address and the date from which it was effective, rather than make a manual note of it and create more work for themselves later.

The system was built with a field for the new address and a field labelled 'effective date'. While it was being tested, Sarah realized that when the 'effective date' arrived, nothing happened. The system did not automatically switch to using the new address. The developers had not correctly understood the need and had not built the functionality the users wanted. Fortunately, as she had planned for an adequate testing and revision stage, Sarah was able to ensure the change to her system was cheap, quick and pain-free – and that in the future, customers would be receiving mail sent to the right address.

Documenting scope is a laborious task, but it is essential to the success of your project that you get it right. That means not only including all the relevant details, but also writing it in a way that is unambiguous. Ursula K. Le Guin gives this advice to aspiring authors in her book *Steering the Craft*:[30]

> **Our standards for writing are higher and more formal than for speaking. They have to be, because when we read, we don't have the speaker's voice and expression and intonation to make half-finished sentences and misused words clear. We have only the words. *They* must be clear.**

Your project scope document is unlikely to be entered for any literary prizes, but Le Guin's advice is still sound for those of us not writing novels. The project scope is not a document for you; it is a list of detailed instructions to someone else. You will not be there to clarify what you really mean when they read it and are working from it, and so you have to make it completely clear.

When you are putting together business requirements for your project:

- Be absolutely certain that there is no ambiguity between what your business users want and your own understanding. Call a meeting and coax out of them what it is they really want from your project. This is an opportunity for complete blue-sky brainstorming – if they could have anything as a result of this piece of work, what would it be? Get them to be clear about their requirements, ask the stupid questions, and then negotiate. It might never be possible to have a dedicated customer service representative for each client or response times to customer calls within two seconds, but those requirements allow you to form an understanding of what is important to your customer: in this example, the business user.
- Once you have the list, add to it, making sure you are being as specific as possible. List colours, materials, brand constraints and any legislation with which the solution must comply. Are there any other systems with which it must integrate? What security is appropriate? Detail exactly what is meant and spell out any assumptions.
- Verify the document with the business team before it goes to anyone else. This is their opportunity to point out that you have not understood what they need. Even better, ask all the relevant parties to sign the requirements document.

  Allowing stakeholders to review the document also means that if the system does not deliver what they were expecting when they first see it, you are in a better position to explain why: you gave them plenty of opportunities to add new requirements or clarify existing requirements. If the stakeholders didn't take those opportunities, they will have to use the formal change control procedure to make any alterations or add new requirements to the project scope at this stage, if it is not too late.
- Once the document is written, test it: give the requirements list to someone who knows nothing about the project and ask them whether they could use it as the instructions to build something. If they think they can, and their solution is similar to what you are expecting, then your requirements document is probably pretty unambiguous. If it is wildly different, then it is time to think again.

HINT

A good requirements document should not only ensure your product is built exactly as you wanted but also form a way of checking off users' needs at the testing stage. It will form the basis of the user-testing documentation, and if it is very detailed it may save you time in writing test scripts.

I DON'T HAVE TIME TO DO ALL THIS

It is inevitable that at some point in your career you will work on a project where there is simply not the luxury of time for a full requirements document. Work with your team to revise things as you go along. It can be risky, but if it is a well thought-out and relatively simple project, this approach can save a lot of time. Just remember to update the requirements document as you go along, so at the end you have an accurate record of what has been done. If you do opt for the make-it-up-as-we-go-along approach, make sure you allow enough time in the plan to test your solution thoroughly, both technically and from a user's perspective. It is much easier to alter things at the testing phase than after the project moves into a full implementation.

Finally, if you are in any doubt about whether your team can handle the ambiguity, take the time to do the requirements gathering and documenting exercise fully – it really will be worth it.

GOLDEN RULES

The users' requirements form the basis and rationale for the entire project, and so it is essential to have them:

- documented;
- unambiguous;
- agreed by the users themselves.

17 Use version control

DEFINITION

Version control is a method of 'recording and monitoring the changes made to a document over time'.[31] It is a simple way of being sure that everyone is reading the same edition of a document and that everyone has the same view of events. It is a discipline, but it is not difficult to do.

THE RIGHT VERSION OF EVENTS

Sue Kettley, an experienced project manager and process-improvement specialist, knows the value of making sure documents are kept up to date and versions tracked. A couple of times in her professional life, she has been able to settle project disputes by having an accurate history of a document's evolution over the lifecycle of a project.

'I was working as a consultant and was called in to help a utilities company sort out a project that was in trouble,' she says. 'They were also working with another consultancy, and that was where the dispute began. The other consultancy claimed the utilities company had specified a particular requirement and then changed their mind. They claimed that the utilities project team had not told them that the requirement was no longer needed, so they had continued to include it in the system build and of course expected to be paid for the time spent,' Kettley explains. 'As their proof, they produced what they said was a working copy of a requirements document that had been given to them at the beginning of the project.'

'Unfortunately, everything had been in a state of flux at the beginning, before my consultancy had been brought in. The document had never been finally agreed. The in-house project team began to panic – they had two consultancies trying to sort out the project, which was in a mess, and they were arguing with critical project team members about functionality that wasn't even needed any longer. And they couldn't prove that they had told the first senior consultant that the requirement had been dropped.

'It was getting really complicated and wasting loads of time. I was trying to be independent and negotiate a resolution between the first consultancy and the utilities company, which was the initial step to bringing the project back in line. I suddenly realized that when we had been brought in, I had created document templates to be used for all the project documents. It was a very simple initiative, but I had put the templates under version control, so I could make sure I tracked any changes to them and monitor how they evolved with the improvements we added.' This proved to be the end to the dispute.

The document the first consultancy were using as 'evidence' that they had been asked to include the requirement was based on Kettley's template. 'As, thanks to version control, I could track back to when that particular document template was created, I found out that it wasn't in existence at the time they claimed to have been given it. End of dispute!' she says.

Kettley spent the dot-com boom years as managing director of a software house in Brighton. 'We were doing joint development projects with our parent company in San Jose,' she explains. The project she recalls most vividly was the development of a new device for fibre-optic telecommunications networks. Kettley set out the objectives of the project in a terms of reference document. She also included a section on issues known at the time. 'I modified the terms of reference as we went through the project setup, and of course each time I changed the document I created a new version and kept a record of the modification history,' she says. One issue relating to the project was of particular concern to Kettley. She needed the parent company to take some action, and as the project went on, as well as attempting to gain resolution in practical ways, she continued to update the issue in the terms of reference document, each time making it clearer that something needed to be done. 'The issue became more strident in each version,' Kettley says. 'Finally, all hell broke loose when it became critical and the US project leader tried to blame one of my team. I wasn't going to allow that – not only had he not reacted to the issue I had continually raised, he hadn't kept previous versions of the terms of reference either. I had!' Thanks to putting her document under version control, Kettley had a record of how many times she had tried to bring the issue to his attention and was able to prove that it was something identified at the very beginning of the project.

You can put any documents – and pretty much anything else, including project deliverables and IT code – under version control. For documentation, you do not need any special database or filing system: just note the details of the version within the document itself. After the table of contents and signoff information, but before the document text really starts, include a table to record:

- the version number;
- the date of this version;
- the author(s);
- the reason for this version and/or a summary of any modifications made.

The numbering is really important and is the key to successful version control. The numbers will be used by your team to check that they have the same version as everyone else. Here are some numbering rules:

- All drafts are written as version 0.1, 0.2 and so on.
- If the document requires signoff, a hard copy of the last 0.x version is physically signed. Then the document is reissued as version 1.0 to everyone on the signatory list. Update the boxes where people signed

by typing in their names and the dates they authorized the document. See Chapter 54 for more on signing off documents.

- If the document does not require signoff, then the version that will be circulated as the 'official' copy is made version 1.0.
- Small changes that result in new versions become 1.1, 1.2 and so on.
- A large rewrite or new round of signatures for a significant re-approval becomes version 2.0.

Have a look at how it works in practice in Figure 17.1 and read through the hints in Table 17.1.

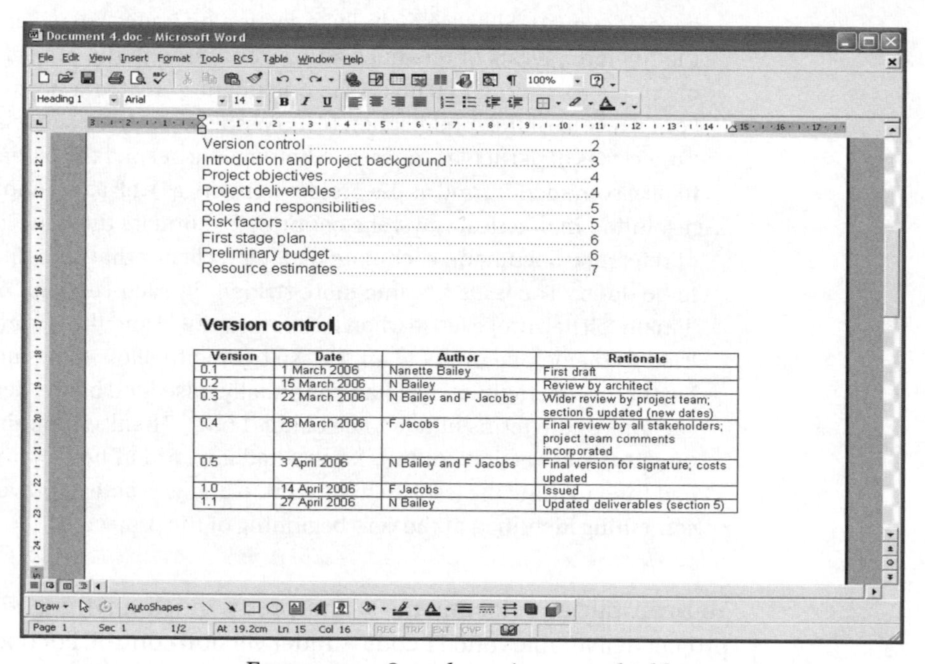

Version control

Version	Date	Author	Rationale
0.1	1 March 2006	Nanette Bailey	First draft
0.2	15 March 2006	N Bailey	Review by architect
0.3	22 March 2006	N Bailey and F Jacobs	Wider review by project team; section 6 updated (new dates)
0.4	28 March 2006	F Jacobs	Final review by all stakeholders; project team comments incorporated
0.5	3 April 2006	N Bailey and F Jacobs	Final version for signature; costs updated
1.0	14 April 2006	F Jacobs	Issued
1.1	27 April 2006	N Bailey	Updated deliverables (section 5)

FIGURE 17.1 *Sample version control table*

If version control is already in use in your organization or on the project but using different rules, then stick to those. Don't get yourself and the team confused by trying to adopt a different standard. The important thing is that you all know what your chosen methodology is and how it works and that you use it without fail.

TABLE 17.1 *Tips for version control*

Do	Don't
Tell everyone on the circulation list when a new issue is released, and send them a copy	Forget to update version numbers in document footers and headers
If this document is referred to in other documents, update those references to reflect the latest version of this document	Issue new documents for every changed comma: if the modifications are of little importance, save them up for the next release
Keep electronic copies of previous versions, preferably in a separate folder called 'Archive versions'	Throw away paper copies covered with handwritten changes until you are sure they are no longer required
Get the right people to approve and review documents first time round to avoid going around the signoff loop again and issuing another version	Hesitate to ask contractors to use the same version-control methodology as your team
Explain to everyone how version control works and why they can't just change documents without you knowing about it and managing it in a controlled way	Try to use version control for a document that changes daily, such as a risk log
Allocate someone (if it will not be you) to manage version control of critical documents	Forget to do it!

GOLDEN RULES

Version control uses minimal effort and will guarantee you always know which is the latest copy of a document.

18 Put a post-project review in scope

DEFINITION

A post-project review (PPR) is a debrief at the end of the project that analyses what went well and what the key challenges were.

The objectives of a review like this are:

- to bring everyone together at the end of the project to formally close it;
- to formalize the key lessons learnt during the project;
- to record this knowledge in such a way that it can be used by other projects in order to avoid the mistakes your project made or to benefit from implementing things that worked especially well.

Many organizations do not routinely carry out PPRs and have no formal way of capturing and sharing project learning. If this is the case in your organization, then it is still very worthwhile putting a PPR in the scope of your project. You and your team can personally benefit both from highlighting the pitfalls encountered in this project to avoid in your next endeavour and from the closure such a review brings.

REVIEWING UNIVERSITY RECRUITMENT

A university in the British Midlands implemented an HR recruitment initiative over a five-year period. The project team and stakeholders had planned no formal review of the initiative. Denise Skinner, a researcher at Oxford Brookes University, interviewed key stakeholders to try to find out management's view on PPRs and why a formal review had not happened for this piece of work.[32] She found that, in principle, the project team felt that evaluating the recruitment initiative was a good idea, but that a formal review had been made pretty much impossible as the project had no definite objectives or success criteria, and so there was nothing against which to compare. 'Those implementing the initiatives perceived that senior management did not place any value on planned, explicit evaluation,' she writes in her paper on the subject in *Human Resource Management Journal*. The team was also concerned that a PPR would end up as divisive and critical, as they perceived the university as a blame culture and this kind of evaluation as inviting negativity.

'The absence of any planned evaluation did not, however, mean that evaluation of the initiative was not taking place,' Skinner continues. Informal evaluations had happened at all levels within the university, resulting in conflicting conclusions. The director of the diversity unit believed that the recruitment initiative mirrored the university's cultural values, and senior management felt the project was successful. 'The unquestioned belief that there would be a benefit to the organization from introducing these particular change initiatives reduced any perceived need to incorporate planned evaluation – those responsible for their initiative already "knew" that their effect would be positive,' writes Skinner.

But lower down the management chain, the view of the project was different. Individuals felt that they lacked the wider vision of the initiative and wanted to have more information about the implementation. 'There was no recognition of the potential, or need, for shared learning or interpretation, particularly outside management circles,' Skinner concludes. 'Individuals at all levels were engaged in making their own assessments and constructing their own "reality" in relation to their experience of the change initiatives. Consequently, future actions were being influenced by assessments based on personal perception and subjective evaluations that resulted from relatively narrow perspectives.'

Including a PPR in the scope of the project will be a step towards making sure the review actually gets done. Add the PPR into your scope statement and the report from the PPR meeting as your project's final deliverable. As the project moves into its final stages, book the date and a room for the meeting, and start asking people to keep the time free. Bear in mind, though, that the fundamental reason for doing a PPR is to ensure that the resulting information is shared and added to your organization's corporate knowledge base in order to avoid 'project amnesia'.[33] This can be done in a very formal, structured way – through a database of lessons learnt, for example – or, as with all things in project management, in an informal way whereby you help the team to reflect on the key messages so that as individuals they will benefit from the experience of the project and take that forward into their next assignments. In reality, you will probably end up being some way along the spectrum of possibilities, producing a report of successes and challenges, which will be circulated to the project team and interested parties.

a

ANECDOTE

Learning from past mistakes could have averted delays and overspends on UK government IT projects over the past 10 years. The Public Accounts Committee has published a good practice guide, which states that it identified problems on over 25 occasions throughout the 1990s. 'Many of the inherent problems had, however, been identified by the committee throughout the 1990s and might have been avoided if action had been taken sooner,' says the 2005 government report into achieving value for money in the delivery of public services. The report says that, despite problems being identified, subsequent projects continue to make the same mistakes, showing that there is a failure to implement lessons learned more widely across public-sector projects.[34]

The PPR should be held at the end of the project and should involve as many as possible of the people who contributed. Invest some effort thinking about what kind of format you want for the review. A meeting where you go through a checklist of questions relating to each phase of the project (see, for example, Table 18.1) may not prove to be the open and honest learning environment you were hoping for, especially if the project hit some serious difficulties. You might decide to interview each individual or team separately, collate and circulate their responses anonymously, and then hold a brief review meeting with everyone in order to agree and share the output. Whatever the format, the same questions will need to be addressed:

- What are you evaluating? Which parts of the project fall into the scope of the PPR? If some parts are excluded, why? Will they be reviewed separately?
- What criteria will the project be assessed against? Can you refer back to the project's stated objectives and success criteria?
- Will you do the PPR yourself? Sometimes it can help to bring in another project manager or an experienced facilitator to run the meeting so you can participate fully. It also makes the environment more neutral, which is especially helpful if you expect sparks to fly between team members.
- In what format will the output be? Are you aiming to produce a report, a memo or a set of database entries? Knowing this will help you structure the meeting to get the best output for you.
- What ground rules do you want to set for the session? It can be useful to specify at the outset that the meeting is not aiming to apportion blame for failures. Agreeing some ground rules can also help (turn off mobile phones, give constructive criticism only, avoid commenting about an individual's performance, depersonalize feedback with language relating to their role in the project or their department and so on). Pin up the ground rules at the beginning of the meeting so they are available for reference as the session goes on.
- Consider getting agreement on whether each point can be documented Be sensitive about what remarks may best be kept within the four walls of your meeting room if the PPR report is going to senior management.

TABLE 18.1 *PPR checklist: example questions to ask during the post-project review*

Initiation/start-up	Were the project objectives, scope and critical success factors identified and agreed? By whom?
	Was a project organization established, with clearly defined roles and responsibilities? Were these documented and signed off?
	Was a cost/benefit analysis or business case drawn up and signed off?
Planning	Was the content of agreed deliverables established?
	Were all products and responsibilities agreed?
	Were dependencies recognized and key milestones identified?
	Was a meaningful schedule created?
	Was a critical path established?
Monitoring	Was progress monitored against the plan?
	Were issues, risks and dependencies recognized and managed in an appropriate way?
	Did the project schedule and budget prove to be realistic and achievable?
	Was effective action taken to address changes to scope, poor quality, time slippage, cost overages and resource issues? How was this done? Did it work?
	Were changes to the schedule and business case appropriately approved and communicated?
	What status reporting was done, and was it accurate?
Structure/ organization	Was the established project organization effective?
	Did the project team function well, sharing views and opinions and owning joint decisions?
	Did the steering committee or project board function effectively?
	Was communication effective at all levels? If not, why?
Outcome/result	Was the project delivered on time and within budget?
	Is the quality of the solution robust enough for the job?
	Does the delivered solution meet the business objectives and detailed requirements?
	Has an operational handover already been done? If not, when is it scheduled for?
	What are the outstanding tasks, and who will take responsibility for them?

GOLDEN RULES

Plan for a PPR at the beginning of your project. Put it in scope and then follow it through, sharing the output with the team and organization for other people to benefit.

19 Identify risks up front

Successful risk management starts at the beginning of a project with the identification of risks: potential happenings that may throw a spanner in the works of your project. This and the following two chapters look at the risk- and issue-management process. This chapter considers the risk-identification phase done at the beginning of a project. Chapter 20 continues the risk-management process by discussing risk response and management. Chapter 21 explains the difference between risks and issues and how issues can also be managed effectively.

a

DRIVING STRAIGHT IN

The lifecycle of the UK government's Ministry of Defence projects includes an initial assessment phase before time, cost and performance targets are agreed and the project is formally approved. The purpose of the assessment phase is to identify and understand the critical risks and put in place plans to manage and mitigate against such risks. The Ministry of Defence's recommendation, based on their past 40 years of experience, is that about 15 per cent of the initial procurement costs should be spent on this phase.

The Support Vehicle Project is a procurement initiative to replace the current fleet of ageing cargo vehicles with new cargo and recovery vehicles and recovery trailers.[35] In March 2001, it was decided to bypass the assessment phase and approve the entire project straight off. The decision was taken because the team believed that the technology was established and there was already enough information about the project. However, skipping this phase meant that critical risks were not identified and the project was significantly delayed. 'Nineteen months of the delay to the Support Vehicle Project are directly attributable to the decision to bypass the assessment phase,' says the Ministry of Defence's Major Projects Report.[36] As a result, the project missed the opportunity to examine risks and plan mitigating actions early on. The tasks had to be done after the project had been given formal agreement to proceed.

Identification [illegible] o note down everything that might [illegible] ving off the project, changes to the [illegible] king longer than planned, changes to [illegible] using new technology, and so on. The project manager alone cannot identify all the potential risks, and so involving the project team will help generate a comprehensive list.

Identify Risk → Analysis

Elkington and Smallman have studied the risk-management process from an academic perspective and arrived at this set of guidelines for risk identification:[37]

- Identify the obvious risks first.
- Think of the who, why, what or when of the project, and identify risks relating to those.
- Consider risks that apply to the management of a project as opposed to the deliverable of your project, such as resources.
- Identify positive as well as negative risks, for example the impact of one task being completed much earlier than expected.[38]
- Use your imagination to cover everything: removing risks from the list is much easier than managing risks you never identified.
- Involve others by working in small groups or holding informal interviews.

Once you have identified as many risks as possible, the next stage is to apply some sense of priority assessment to each item. You will have a very varied list of risks, some of which will seem small and pointless, but others very significant. There are two attributes that should be considered for each risk: likelihood and impact.

- **Likelihood:** the chance of the risk occurring. Unless it is a very scientific event, you'll have to take a best guess based on your gut feel about this.
- **Impact:** how serious the outcome of the risk would be if it materialized. The impact could be to the project schedule, budget or the quality of deliverables. Again, you will probably find yourself being less than scientific when it comes to assessing this.

Each of the two attributes is given a value, and these values are multiplied together to calculate the overall risk assessment. 'Research shows that the severity of the potential consequences of a risk produces a greater concern than its probability in evaluating the overall level of risk,' write Baccarini, Salm and Love in *Industrial Management and Data Systems*. 'For example, a low-probability/high-consequence risk is typically considered as being higher than a high-probability/low-consequence risk.'[39]

Figure 19.1 shows a risk matrix and two examples of risk assessment in practice. The first example is the lack of resources for user testing. There's a possibility that users would not be available to carry out testing when needed, as they all have operational priorities. This has been assessed as 4 on the risk likelihood scale. The impact on the project timescales would be moderate (assessed as 3).

The likelihood of facing an office flood is tiny. This has been assessed as 1 on the matrix. The impact on the project if the office was flooded out would be severe, and that has been rated as 5.

Combining the likelihood and impact scores gives each risk a relative priority. User availability for testing ($4 \times 3=12$) is a more significant risk than flooding ($1 \times 5=5$). Once each risk on your list has been assessed and given a score based on its priority, you will be in a position to consider how best to manage it, which is discussed in the next chapter.

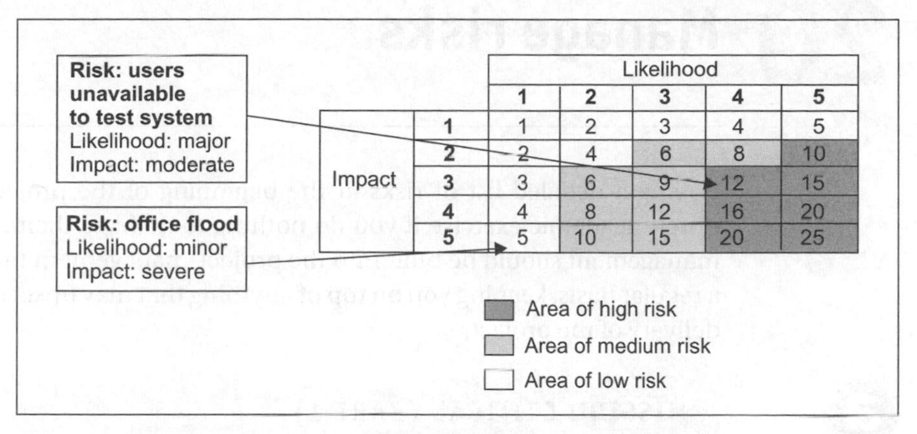

FIGURE 19.1 *Matrix for calculating risk priority*

GOLDEN RULES

Identification should be done early in the project and involve the project team. Each risk should be assessed according to impact and likelihood.

20 Manage risks

Having a detailed list of risks at the beginning of the project becomes a purely academic exercise if you do nothing to manage them. Ongoing risk management should be built into the project-management tasks you do on a regular basis, keeping you on top of anything that may upset the successful delivery of the project.

MISSION CRITICAL (PART 1)

Dr Ady James, a project manager with 14 years experience in the management of space projects, led an international team of around 30 people to develop a highly technical extreme-ultraviolet imaging spectrometer. The instrument forms part of a Japanese spacecraft that has been designed to study the sun.

The project started in 1998 as a collaborative effort between the UK, the USA, Norway and Japan, with a budget of around £13 million. Managing the risks associated with such a large and critical piece of work was essential.

'The main aspect of our risk-management system is that it is paper-based and under the control of the project manager,' James explains. 'It requires very little maintenance other than the vigilance of the project manager.' Deceptively simple, the project team was constructed in such a way as to make risk management foolproof. James' team, responsible for the spectrometer, tracked risks relating only to that. 'The spacecraft-level team had their own risk assessment, and so we would be tracking the same risks. Similarly, the local teams had local management and would be tracking their risks. My risk identification therefore passed some risks up and some down, and I received risks from both directions. This acts as a failsafe for spacecraft-level risks. If we didn't identify it as a risk, then someone else might.'

Near the start of the project, the whole project team was asked to identify potential risks. These were then added to a draft risk-assessment document. A small group of the project team with responsibility for the delivery of the technical elements of the project reviewed the document and categorized the risks. Risks were divided into two major sections to help with the classification:

- programmatic risks, associated with the build phase of the instrument up to launch;
- operational risks, associated with the operations phase of the mission, i.e. post-launch.

Then they were split further into another two sections:

- system-level risks, i.e. risks defined as having impact beyond the experiment and that may affect the spacecraft or mission;
- subsystem-level risks, i.e. risks contained within the experiment and therefore likely to impact only performance.

'This is where we split away from current thinking on risk management,' James says. 'Normally, one would look at probability of occurrence and the impact if the risk was realized. Generally, any impact on performance will be unacceptable to the end user, the scientist.'

There were performance tolerances for the spectrometer, and the role of the risk-management exercise was to resolve any risk that put the instrument outside of this tolerance. For that reason, the risk impact was documented but not scored – every impact was treated in the same way.

'Another split from traditional thinking is that, rather than an individual, we name the local manager as the owner of the risk as someone who can locally delegate,' James says. 'It is up to the local project managers to report to me on risk and assign an owner internally if appropriate.'

The project team updated the risk register and reviewed the list of actions during their regular meetings, and at the end of significant phases the register was reviewed by a panel. Occasionally, new risks were added at this point, and the panel had to be satisfied that all risks were being managed adequately before allowing the team to move on to the next phase.

James had originally planned to update the register with new risks on a monthly basis, but in practice there were not that many to add. They tended to be minor performance concerns or to relate to unplanned work. 'I found that the frequency of controlling all of the management tools was driven more by the needs of the project than by any active plan on my part,' he says. 'When things were going as well as expected, a dogmatic adherence to updating risk registers and the actions list and chasing the individuals for the sake of an update did not seem productive.' James believes project managers can get a feel for this from management by 'walking around'. 'You get a feel for the team's concerns and know when you may want to formalize some of the control functions,' he explains. 'As the team matures into the project – many of mine were already experienced – I found that they were all so aware of the needs of the project that management was always more of an oversight activity rather than a forced driving-type activity. I have the advantage that nearly all of the team members are suitably motivated by the very nature of the work, but I suspect that not many industries get that for free.'

The majority of the risks related to the fact that the project was using untested technology. 'The number of risks in this type of project reaches a plateau very early on, and additions are rare,' James says. 'In our domain, we find that risk management rapidly integrates with the project scope, rather than being seen as a separate activity, so updating the risk register is not so essential. The level of risks at the subsystem level is also generally low as the result of a fairly conservative industry and an experienced development team.'

James kept the register tidy by closing down risks that had passed and not materialized, which were mainly those relating to late deliveries or test failures. 'Despite the low number of issued updates to the document, it was read by me at least on about a monthly basis,' he says. 'I set up reminders on my calendar software to remind me to check and update the risks, as well as the budgets, schedules, actions and issues, on a monthly basis.' James' advice to other project managers is to never plan to do these updates on a Friday. 'Don't plan to do them all on the same day either,' he adds, 'because it won't happen. You never get a full free day when a project is in full flow.'

Risk management can be done in an incredibly formal way, with scheduled meetings where the team gets together to discuss the latest developments for each risk, or in a less structured way, with the project manager coordinating the management activity and getting updates on risks as part of the normal progress updates from the team. How you choose to record your risks is up to you; a starter-for-10 document template is included in the appendix.[40] The point of having done the identification exercise is to then put plans in place so the risks never happen.

Whatever your approach to the paperwork side of things, your approach at the business end of risk management is going to be the same: work out how to handle each risk, plan actions to meet that strategy and monitor progress against the actions. This is the actual activity relating to risk management as opposed to the documentation and thinking process that has to happen up front. The activity of risk management is the critical part, as studies have shown that having an up-to-date risk-management plan and a process for assigning ownership of risk statistically improves your chance of completing the project on time.[41] The more risk management you do, the more chance you have of successful project delivery, however your project defines success. It is hardly surprising that research shows that the riskier the project, the less successful the outcome; some projects, however much risk management you do and however many times you allocate actions or chase your team for updates, are just too risky to really deliver a top result.[42]

The next step after risk identification is to work out what you are going to do to stop those risks from happening. This is called risk response. There are four general risk response strategies, as shown in Table 20.1.

Baccarini, Salm and Love have looked at the popularity of each of these different types of risk response in a study of Australian IT project managers. Presented with a list of frequently occurring risks that were not specific to any particular project, such as lack of resources, incomplete requirements, unreasonable project schedules and so on, the project managers described the strategy they would apply to managing the risk. The results were incredibly varied. The authors conclude that this 'indicates that there is not one solution for managing any particular risk and the project manager must be aware of the possible need to implement two or more treatments for one risk.'[43]

TABLE 20.1 *Risk responses*

Response	Description	Example response: risk – bad weather may spoil the school fete
Avoidance	Refrain from carrying out the activity that will result in the risk occurring	Don't hold the fete
Reduction	Act to reduce the impact of the risk should it occur or the likelihood of it occurring	Hire marquees for all the stands, so the fete can go ahead under cover if the weather is bad
Transference	Get someone else to take on part or all of the risk	Take out an insurance policy against the potential loss of income for the school if bad weather keeps people away
Acceptance	Understand and accept the consequences should the risk happen	Accept that there is a chance of bad weather and do nothing

As their research shows, there is no textbook way to manage any given risk. What works for one project in one situation may not work for exactly the same risk in a different project and a different situation. You will need to use your judgement to decide what action plan will be the most effective for you. Baccarini, Salm and Love's analysis does show that reduction is the most favoured risk response; transfer and acceptance are hardly used at all. This may be because in the majority of projects, it is difficult, if not impossible, to transfer the risk to a third-party contractor, an insurance company or even another department willing to take the chance. Acceptance is similarly not proposed frequently as a risk response, because it is suitable only for very small risks. It is rarely appropriate to do nothing.

Once you have decided on the right risk response given the situation, you can work out a risk-mitigation plan. This is a task-based approach with the aim of making sure that the risk does not materialize. Draw up a list of tasks and owners for each action to carry out your risk-mitigation plan.

HINT

Your risk-management plans can become a formal part of the signoff process to move from one phase of the project to another. It's not a good idea to talk to your sponsor about the project risks only when you move from one phase to another. They will probably want greater visibility of the risks facing the project, so you should provide that information on a more regular basis. One way to do this is to include the top three risks of the moment in your regular reports to the project board, along with a brief statement of how you are managing them to mitigate against the potential impact they would have on the project.

A final thought on risk management: once you have successfully mitigated against a risk to the point where the risk will now not happen, it can be removed from the risk register. That does not mean deleting it from the document altogether. Each risk should have a status: open or closed. Move closed risks to an appendix or somewhere at the bottom of the document out of the way. Keep a record of what you did to manage the risks out of the project: this might be useful for your post-project review, for an audit or if a similar risk comes up later in the project.

This has been a brief introduction to risk management. Whole books have been written on the subject, so see the further reading for some ideas to follow this up.

GOLDEN RULES

Having a list of risks is not enough. Risks should be managed by defining the appropriate risk response, allocating an owner and carrying out activities to mitigate against the risk materializing.

21 Manage issues

DEFINITION

An issue is a problem that is affecting your ability to deliver the project successfully. It might be big or small, something that can be fixed in a day or approached with a long-term vision, but all issues will be handled in the same way.

MISSION CRITICAL (PART 2)

Dr Ady James' spectrometer project did not have an issue-management system in place at the outset in 1998. He adopted an informal approach to documenting issues, combined with a close knowledge of his team's activities and concerns to ensure they were managed effectively.

'The issues register was informal and anything could go in it,' he says. 'I would cut and paste sentences from emails from concerned engineers and give it an issue number so it could be tracked.' James led his international team from the Mullard Space Science Laboratory at University College, London, and engineers from the UK, the USA, Norway and Japan contributed to the issue list. James noticed that the issues raised fell into two categories: everyday concerns and insecurities, and new items that had not been conceived in the project planning. 'The issues list was a way of allowing anybody to input new tasks into the project plan,' he says. 'Unplanned work issues or points relating to practical build or assembly were raised as issues in the issues register, where they were turned into new work packages and therefore planned tasks or actions.'

With a budget of around £13 million, partly funded by the Particle Physics and Astrophysics Research Council, it was essential that every concern was addressed. The issues register was one of the ways James could demonstrate to his team and stakeholders that things were being managed. Like the way the project managed its risks, the system was very simple and based on the circulation of documents rather than any complex IT package. James is clear that expensive issue-management applications are not necessary: 'I like the paper-based systems in that I only have to worry about the update and maintenance of the information and not the update and maintenance of the system that contains the information as well,' he explains. 'Your project tools should support your project and no more. There is a fashion for believing that the quality of the project is reflected in the quality or complexity or newness of the tools used or in how the information is displayed. I don't believe this. Some tools will be necessary when the size of the project is such that a paper system is unworkable, but if it is not needed, don't use it. The management of the project should be challenging enough as it is. When there are major problems, teams like to see that the right decisions are being made, actions allocated, resources freed up. The use of these tools, whether paper-based or an IT system, then becomes a good way to show how we would get over the issue,' he says.

Issue management requires the same approach as risk management: log the issue, devise an action plan, carry it out and monitor the situation. The issue log should include any problems that occur that will have an impact on your ability to deliver the project successfully. An example log is included in the appendix.

Issues may be the result of a risk that has materialized but could just as easily be something that has never been on your risk register. The log can include issues over which you have no direct control as well as those you can fix using the resources within your project organization. However, this is not the place to log approved changes. They should be kept separate, so if the result of an issue is to raise a change to address something, the change should always go through the change-management process.

By their nature, issues are more immediate than risks, and as such you may choose a more informal approach to documenting them. As the situation has already happened, it is less useful or necessary to craft a descriptive paragraph explaining the issue: a cut-and-paste from an email may well be sufficient to provide enough of a memory jog so that your team knows exactly what is being referred to.

The appropriate resolution for an issue may be immediately obvious, so that you see quickly what you need to do in order to create an action plan to rectify the situation. That's rare though. It is more likely that the full extent of the problem will not be known. Then it is down to you to investigate the issue thoroughly, drawing on the relevant team members, in order to be able to find an appropriate solution.

A good way to analyse a problem's root causes is to brainstorm with the right people in a room. This will help you:

- understand the issue;
- be able to explain its impact to the relevant stakeholders;
- devise an action plan with which you can monitor the situation.

Due to the immediacy of issues, there are two key things to bear in mind:

- the efficacy of the action plan to handle the problem;
- the speed with which you allocate someone as the owner of that issue to take responsibility for seeing that the action plan is carried out.

Even issues outside of the project's control can have a team member allocated a watching brief, someone tasked to provide updates on a regular basis as an issue unfolds. It may take a significant time for some issues either to be resolved or to disappear. On the other hand, some issues may be a short sharp shock to your project, over and done with very quickly.

HINT

Issues can be controlled in various ways depending on the type of problem, but a common way to resolve simple issues is to propose a change to the project schedule, budget or requirements in order to incorporate the tasks needed to resolve the problem. If the project sponsor agrees, then the change is approved and you can replan your project accordingly, proceeding with the new status quo.

The issue status in the log can be changed to 'closed' once an issue is controlled. As with risks, do not delete issues from the log. The complete list of issues with the action plans and final resolution provides a useful audit trail and input into the post-project review. It will also help you remember what you did if you come across a similar problem later.

GOLDEN RULES

Issues have already happened, so log them, draw up an action plan to manage them and move quickly to execute that plan.

22 Document assumptions

In any project, there are things you don't know. It is impractical to wait until everything is a known fact before work on the project starts. In order to start work, it is necessary to form some assumptions – statements about what you believe to be the case – to create a position from which to begin. These will be proved or disproved as you work on the project.

WORKING IN THE DARK

'One killer in terms of scope and budget for projects can be assumptions,' says Neville Turbit, a convenor of the Australian Computer Society Project Management group and principal of the project management software and consultancy firm, Project Perfect.

He spent some time working with an Australian government department to help them implement a project-management methodology, which included handling assumptions in a pragmatic and practical way. Teams often find themselves in the dark at the beginning of a project, which means assumptions form a necessary part of the foundations of the requirements and plans. 'I talk to my teams about travelling down a tunnel with only a flashlight. We can see clearly what is immediately in front of us, but only some of what is down the tunnel,' Turbit says. 'We know there is more down the tunnel than we can see. We just don't know what it is. That's the way it is with scope. We know there is more scope than we can see. We just don't know what it is until we get further down the project tunnel.'

It is the same with assumptions about budget and time. 'At first I started trying to get them to add a contingency to the budget. And I tried to get them to leave a hole in the Gantt chart for the unforeseen. There was so much resistance to unallocated resources that this proved almost impossible. The line management way of thinking does not fit with a project environment and its so many unknowns.'

The government department learnt to handle assumptions as ongoing open issues, with actions required to validate them. Turbit gives an example: 'If we make an assumption that we will not have a distributed database, then the project team immediately creates an action to validate the fact that it will not be a distributed database. The action is assigned to a person, and given a date for completion. That action is then monitored by the project manager.' This means that assumptions don't stay assumptions for very long. Either the result of the investigation is that the assumption is true and it becomes a known fact, which aids the further development of the project. Or it is false. 'If it proves an incorrect assumption,' Turbit says, 'we call a meeting to look at its impact and what other actions may be required to compensate. Too often in projects I have seen assumptions documented and forgotten,' he says.

DEFINITION

Assumptions can be:

- things that you are taking for granted will stay the same;
- things you have to assume because you don't yet know for sure.

In the first case, it is worth documenting these assumptions in your project initiation document, even if everyone accepts that is the way things are. Just because something is like that now does not mean it will stay that way. The way organizations work changes more often than a project manager can keep track of: a change in an area where you least expect it can, and will, interrupt your plans. Examples of this type of assumption are:

- Payment for contractors will stay at £900 per day.
- The marketing department will make resources available for testing.
- The October 2006 IT security standards will be used to develop the application.

If payment rates change, marketing cannot provide testers or the security standards are updated, you suddenly face a very different project – one that you probably will not have the budget or time to complete. You will still be the project manager and be expected to deliver but, having documented the assumptions, it will be easier to explain to your sponsor why you now need more money, time or people. The sponsor originally agreed to a project in a certain environment: now that has changed, with all the relevant knock-on impacts to the project budget, scope and timescale.

The second type of assumption will allow you to plan more accurately. You might not yet know the number of staff that will need to be trained on the new accounting system, but you can assume it will be half the finance department, say 35 people. Having stated and documented this, you can decide that they will be trained in two groups and book those dates, plan the costs for the trainer, hire the room and so on. Planning based on assumptions almost always results in a potential issue, so add an issue to your log: 'Number of delegates to be trained still undefined.' Allocate an owner to the issue and make them responsible for validating the number of staff that need to go through the training. Once you have a concrete answer, you can replan as appropriate.

WARNING

A word of warning: use assumptions with care! It is always better to have the full picture and work with concrete facts. Making assumptions such as 'I will have a budget of £500,000 to set up the company football team' will not help you manage the project at all. The fewer assumptions you have, the more likely it is your project will avoid surprises later on and the more realistic your plan will be.

GOLDEN RULES

Document all project assumptions in the project initiation document. Document and monitor the associated issues and update your plan when you have validated each assumption.

23 Involve users in scope definition

Project communication works in two ways. This chapter looks at one of the methods to get information from your team and relevant parties. The following chapter considers how to give information to those key stakeholders. Involving users in defining the project scope is fundamental to the success of any project. You can't define the scope sufficiently alone by brainstorming what you think the users really want the project to cover. Although some project managers do take this approach, there is a risk of both alienating your users and missing something important. Including your customers, the end users, in this part of the project definition will guarantee your scope is as comprehensive as possible. Your role is to extract all the necessary information from them and guide them towards defining a useful statement of project scope.

a

DOT-COM TELECOMS (PART 1)

During the dot-com boom of 2001–2002, Liz Kirby was involved in a £6 million programme of activity to consolidate the web and intranet sites from 81 countries into global ecommerce presence for a major telecoms company. She led a 15-strong team on the intranet content workstream. The task was to consolidate six main intranet sites plus a host of country-, function- and product-specific sites running niche applications. She found that a lot of the small sites had fallen into disrepair as they lacked the support to keep them running.

'We wanted to have a single content management system, a single search engine, one set of processes, and the organization to support it all,' Kirby explains. 'The project had implications for people's jobs – the intranet was going to be managed by a new UK team, so people in the territories were writing themselves out of a job.'

Consultants were brought in to manage the first 12 months of the project, during which time the site design was drawn up, the UK team recruited and the new search engine and content management system bought and implemented.

'There were no "business" people on the team,' Kirby says. 'They were engaged when something needed to be signed off, usually because the project team had come up with a list of things and needed the business to rank them in order of importance.'

Business user involvement was complicated further by the fact that the IT department owned the relationship with the third party that had provided the content management system, but the system was used primarily by business people and not by the IT department.

'It was incredibly detrimental to the project team not to have a relationship with the vendor until later on,' Kirby says. 'We were literally given a box with high-level instructions on the side, and a copy of the content management system was made available on a server in the USA. From that, we had to deduce how best to make it work. It was equivalent to buying a complex piece of software like SAP and having no specialist consultancy to explain how to get the best out of the service. It was only much later on when the project team forged a relationship with the vendor that things really started moving, and we could understand how the software might actually deliver what we wanted from our intranet. We had made lots of assumptions about the technology, which were simply incorrect, but totally understandable given the situation we found ourselves in – nobody from the vendor was there to match up our requirements to the functionality.'

As users were not involved in the project definition early on, their needs were largely ignored. 'The scope was documented in a haphazard way that didn't relate to how users would end up interacting with the new intranet,' she explains. The focus was very much on ensuring the sponsor's requirements were met. What the sponsor wants is not always the same as what the end users have asked for.

The project was delivered on time and under budget, and it met the sponsor's requirements: get costs down and have a process for publishing global internal information. 'But we missed what success meant for the end user,' Kirby says. 'There was no post-project review, no analysis of whether it helped people do their job more efficiently.'

Liz Kirby is clear about what she has taken from the experience of working on this two-and-a-half-year project. 'There's no point in doing anything without business involvement and a really clear understanding of the business requirements. It's impossible to do adequate testing or a post-project review without getting users involved in defining the project scope up front.'

During the scoping stage, your objective is to define what the project will encompass. It is not the place to get bogged down listing technical requirements and the colour of the wallpaper in the new office building. The scope of the project is what will be covered or touched by the work. Where does the project start and finish? Consider the following:

- Be clear. As with all project documentation, use the most specific words possible. If in doubt, clarify. Does 'launch' a new intranet mean design it, buy the hardware and software, build the technical infrastructure and switch the site on? Or does it include producing training material, running courses on how to use it for every department and writing communication material?
- Document what is not included in the project. If you focus only on changing the employment contracts for staff in sales, say so. 'Updating

the contracts for all other staff' should be listed in a section of the document titled 'out of scope'.

- Clarify that you have understood what your sponsor wanted to achieve by the project. Book a meeting with the sponsor and any other key stakeholders with the objective of verifying the scope statement. Send them the document in advance: they might have time to read it and reflect, coming to the meeting better prepared, but plan to walk them through it anyway as the reality is that many stakeholders won't make time to read papers sent out before a meeting.

Plenty of research shows that involving users has a direct relationship with how satisfied those users are at the end of the project.[44] Getting them involved is not always straightforward. The sponsor will often see it as your responsibility to define and document scope and will need convincing of the value of involving the end users at this stage. However, projects are done for (and sometimes to) the end users. Project managers have the luxury of moving on at the end of the project and not having to live with the consequences.

Having said that it is not a good idea to pull together a scope statement by yourself, it is a good way to get started. Call a meeting of all the relevant parties and present them a 'straw man': the draft scope statement. The objective of the meeting is to get commitment and input from those people who will be touched directly by the project. Ask the question: 'If we did this, what would be left out? What would we do that is really unnecessary?' Be prepared to offer some suggestions to start them thinking. A good exercise is to pair up those present and ask each pair to try to break down a phrase in the scope statement into at least three more specific phrases. Then ask everyone to share and discuss and, crucially, agree their results. By the end of the meeting, you should have a much clearer idea of what your stakeholders actually want from the project.

HINT

Handwrite your scope statement on flip-chart paper or an overhead projector acetate. You want it to have the air of a work in progress. Giving each attendee a copy of a professional-looking document, properly typed up, will discourage people from suggesting changes: the scope statement already looks finished, and it takes a vocal team member to volunteer how it could be improved. This is one time where sloppy presentation is a plus.

AN EXERCISE FOR EXPLAINING THE IMPORTANCE OF FULLY DETAILING REQUIREMENTS

If your users see defining requirements as straightforward and the process of documenting them a waste of time, then they are probably not thinking deeply enough about what the project entails. This simple exercise can be used at the beginning of a requirements-gathering workshop to explain why it is important to write down each requirement in what can seem a painful level of detail.

(i) Ask the group how difficult it would be to program a robot to cross the road at a set of traffic lights, and gauge their reaction.

(ii) Ask how long it would take to come up with the sequence of decisions and actions they would use to program the robot.

(iii) They will probably suggest a time in minutes for this simple task. Give them as long as they have asked for to actually do it, but no longer than 20 minutes. Split them into pairs and ask them to write down a list of decisions and actions to get their robot to cross the road.

(iv) At the end of the time, ask each pair to report back and encourage the other pairs to critique their list.

(v) Inevitably, the robot will:

- fall off the pavement because it forgot to step down;
- get hit by a bus that didn't stop at the light;
- walk over the person in front of it;
- walk into an oncoming person;
- get wet, because it is raining, and short-circuit;
- not know when it has reached the other side and trip over the kerb;
- not be programmed to cope with snow;
- and so on.

Do the exercise yourself beforehand and come up with as many conditions as possible that you think your user group might not consider.

This exercise is usually enough to get users to understand how complex requirements can (and should) actually be. You can then move on to discuss the project requirements themselves, encouraging the group to be as specific as possible to avoid a malfunctioning robot.[45]

Another reason to get users involved is that it minimizes the mental model mismatch (see Figure 23.1). This term refers to the syndrome of users not getting what they want because what they described at the outset was transformed in the minds of those who actually then developed it – by the time it got back to the end user, it was subtly different from what was originally requested. Every person involved changes the end result just a little, as they have their own interpretation of what that end result looks like.

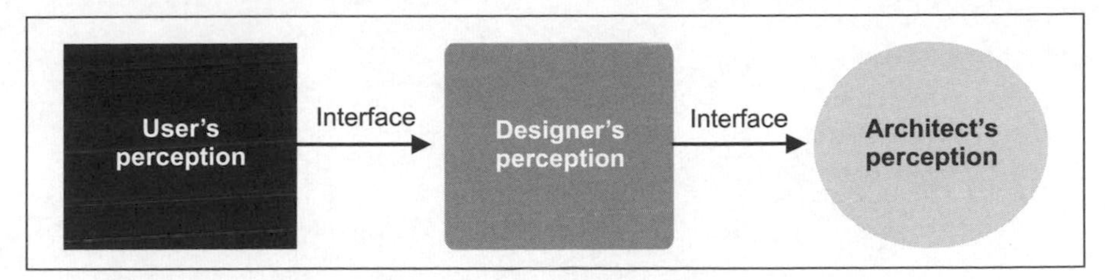

FIGURE 23.1 *The mental model mismatch*

The mental model mismatch happens for two reasons:

- The requirements are not explained well enough, so what the end user thinks they have described and what the person recording the requirements thinks they have described are slightly different.
- Those tasked with delivering to the requirements feel they can improve on the original ideas slightly, so they change things just a little as they go along.

Getting users involved in this early stage of the project will counter both these possible causes of the mental model mismatch. First, they will have plenty of opportunities to describe exactly what they want and be part of the documentation process. Second, the project team that will end up building the new IT system, or whatever the project will deliver, will have a chance to meet users before it gets to the testing phase. They will really be able to understand what the end users want and to appreciate that just because a new tweak is possible doesn't mean that it should automatically make it into the final deliverable unless the users agree that it really is an improvement.

GOLDEN RULES

Include as much detail as possible, with input from your users and stakeholders, to ensure you do not leave anything out of the project scope statement.

Section 3
Managing project teams

The control of a large force is the same principle as the control of a few men: it is merely a question of dividing up their numbers.

Sun Tzu (6th century BC), *The Art of War*

Project management is about getting people with the right mix of skills together at the right time to deliver something in response to an identified need. A project usually involves a group of people whose day jobs have very little in common but who have been thrown together to create a project team. As the manager of that team, you will probably not have line-management authority over the people working with you, but you will still need to instruct, motivate, coach and cajole them into producing accurate estimates, progress reports and the project deliverables without taking too long or spending too much money.

Matrix management is not an easy skill, and handling the relationships within your team is what this section is all about. It covers managing your team, sponsor and project stakeholders, as well as soft skills in general.

24 Communicate and document changes

Despite having a clear and precise idea of what your project will deliver, agreed to by all parties at the beginning of the project, it is inevitable that there will be changes. When this happens, you need to be certain that you take the opportunity to explain the new changes to everyone concerned. Communicating **to** stakeholders needs a different set of skills compared with extracting information **from** them.

DOT-COM TELECOMS (PART 2)

Liz Kirby's project to bring intranet sites from four corners of the world into one consolidated system suffered from another problem: the scope was not written down.

'It was difficult to keep a grasp of the scope, as it kept growing' she says. 'If the business were prepared to throw money at a problem, the scope just expanded to include the problem. We had written a terms of reference document and thought there was a common vision, but everyone still had different interpretations of what was supposed to happen.'

This inconsistency in the scope affected the project communications. It was hard to make it clear to the end users what the 'launch' of the new intranet site actually meant. The inconsistency also affected the way in which IT and the intranet teams around the world interpreted their parts in the project. Communications issues came to a head eight months into the project, when Kirby realized that she had to act to ensure everyone had the same message. She sent out members of her project team to visit regional offices around the world. She went out to Japan and spent a few weeks with the intranet team there to clarify exactly what the project would deliver and what she needed them to do.

As well as validating the project progress, this approach had some other spin-off benefits. 'With a clear scope, you can trust people to go off and do what needs to be done without having to monitor the detail,' Kirby says. As her project team was only 15 people, it was essential that she could trust the local teams. 'It was also easier to resolve conflicts later on as we had built up relationships with the people involved around the world. And as they had not worked on anything like this before, it was a useful learning experience for the team. They can take that with them on to future projects,' she adds. 'It was difficult to manage the in-project communications over the two-and-a-half-year project because of the scope changes,' Kirby says, 'but harder to manage the messages to people outside the project. If the messages change, you just don't look credible, and that doesn't help anyone.'

Communication during any project is important, but when elements of the project are changing on an apparently random but regular basis, it becomes critical to get across the right message in the right way. Goodman and Truss studied two major change initiatives – one at a company undergoing significant restructuring and the other at a company moving offices. Very few managers or employees felt that the communication around these projects was adequate. Goodman and Truss present a 'best approach' model developed from their academic research as well as the experiences at the two organizations they studied.[46] This model, the change communication wheel, is shown in Figure 24.1. Although it was designed to help plan communication strategies about change programmes as a whole, i.e. to a very wide audience, it is also relevant to in-project communications and the communications to your project team.

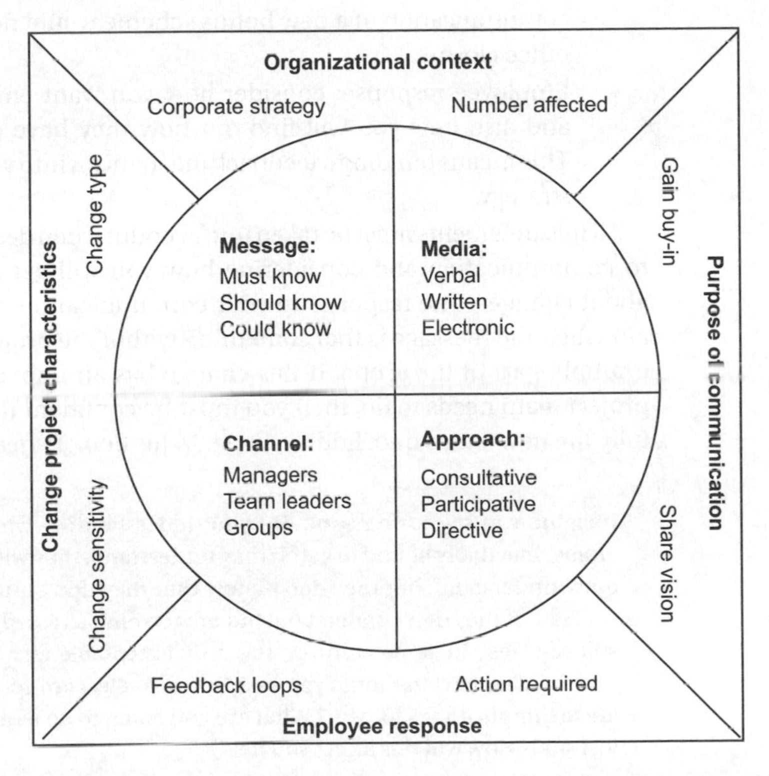

The change communication wheel, adpated from Goodman, J. and Truss, C. (2004). The medium and the message: communication effectively during a major change initiative. *Journal of Change Management*, 4:217, No. 28.

FIGURE 24.1 *The change communication wheel*

The model illustrates the four elements of communication where a decision is required from the project manager:

- What is the message?
- How will we present it?

- How will people hear about it?
- What is our strategic approach?

The correct response to each of these questions depends on four external factors:

- Organizational context: your decisions will depend upon the situation within your organization, as what would work in one company may be disastrous in another.
- Purpose of communication: your responses will depend on the stage you are at in the project, as the aim of your communications will differ as the project progresses and the audience reacts to previous messages.
- Change project characteristics: different strategies and decisions are required for different types of project. What would be suitable for the communication of a new bonus scheme would not be relevant for an office closure.
- Employee response: consider how you want employees to respond and also how you will find out how they have actually responded. This means building a feedback mechanism into your communication strategy.

All these elements must be taken into account when designing your approach to communication and considering how you will get across the message about changes. The response to your communication is especially important when the message is that something within the project has changed, for example part of the scope. If this change has an impact on the work your project team needs to do, then you must be confident that they understand fully the new status quo. Eddie Obeng, in his book *Perfect Projects*, writes:[47]

> **Imagine you say to someone 'Do you understand?' – what answer are you almost inevitably bound to get? If they understand, they will reply 'Yes'. If they don't understand, but they don't know that they don't understand, they will say 'Yes'. If they don't understand but are too embarrassed to say so, they will still say 'Yes'! In some cultures, the only acceptable answer anyway is 'Yes' . . . So what question should you ask? One of a select group. Ask instead, 'What are the implications for you? What are you going to do as a result of what I've just said? How will this affect you next?**

This approach gives you the confidence that they have understood and taken on board the change. It has the added advantage of offering you the opportunity to correct any misunderstanding at this point and not three weeks later when they deliver something completely different to what you were expecting.

HINT

When sending emails, use a descriptive subject line, not 'January Project Report' or 'Re: Your Question'. Even if the recipient does not have time to open the email and properly digest the contents, they can at least get an impression of the situation from subject lines such as 'Project Alpha delayed by two weeks' or 'Project Beta passed audit'.

There is the additional factor of maintaining credibility when you have changes to scope or other elements within the lifecycle of the project. Trust is an important factor here, and when your project team is split over multiple locations, even within the same building, maintaining a level of trust within the team will allow you to maintain the project's credibility. It is far easier to write 'build trust within the team' than it is to explain how to start doing this, as trust is something that develops over time. Honesty and predictability – doing what you said you would when you said you would and behaving in a way your colleagues would expect – are two factors that will provide a starting point for you to work out what trust means for your team.[48]

Finally, avoid a one-size-fits-all approach to communicating change. Your project team will need a different level of detail to your sponsor. Equally, those affected by the change will need a different message to the one you give your boss – not different factually, as inconsistent messages will damage your project's credibility, but presented differently and with a different amount of detail. Make the time to give the detailed version to anyone who asks, but tailor your communication to suit the needs of your audience.

GOLDEN RULES

Projects change the status quo, and projects themselves change as they progress. Effective communication is essential at all points through the project lifecycle in order to ensure understanding and maintain credibility.

25 Get them to believe

It is always easier to inspire your team about the end goals at the beginning of any project (although obviously it depends on the project). The messages you offer at the outset will set the tone for the subsequent reinforcement of those messages throughout the project's lifecycle. Commitment and belief in the project's objectives are seen by some project leaders as nice-to-have extras, but you will find it a lot easier and more pleasant to work with a dedicated team. With that in mind, start as you mean to go on and aim to inspire the team from the outset and throughout the project.

INSPIRATIONAL INDIA

Six months after the Nectar loyalty card was launched in September 2002, Loyalty Management UK was already looking at how to make the card more successful. Collectively, the company's management team has over 100 years' experience in the loyalty market, so they had plenty of ideas. They decided to introduce an online service for cardholders to redeem the points they had collected through the scheme's various partners, including Sainsbury's, Debenhams and Barclaycard. The online service would also mean cost savings at the contact centre, which was dealing with queries generated from having 20 million cards in circulation.

IT director Mark Brighton knew that the 50-plus project team would need to be fully committed to the project to ensure the work was completed within the 12-month plan. This was always going to be a challenge, as a separate team was formed in India to complete much of the development work. 'We went out to India to see the offshore team to explain our vision,' Brighton said in an interview for *Computing* magazine, 'and we used videoconferencing to give them updates on the company strategy.'[49] Brighton's opinion is that companies need a real commitment before they make the move into offshore development. That the Nectar team was '150 per cent committed' can be seen from the results. The Nectar online project delivered on time and on budget and achieved a cost reduction in the contact centre of 26 per cent.

WARNING

In the early days of your project, you will probably find your team hugely committed to achieving the objectives. But there's a risk this will wane as soon as their day jobs start to take precedence over project tasks. Realization that the work cannot be completed instantaneously will set in quickly. As the project manager, you have the responsibility of keeping your team focused on, and dedicated to, the job in hand. This is a task made more challenging by the fact that some of your team may not have chosen to work on the project willingly and are just doing what they have been told. Making them believe in the long-term success of the project is the secret to ensuring a good working relationship with your team and in achieving your end goal.

Take time to explain the goals of the project: do not assume that just because your team member's boss has allocated them to the work he has also explained the ultimate objectives. You know why the project is important, so translate that into reasons why it is important for the people working on it. Put the project into the context of the company's overall strategy so individuals can see the link to the bigger picture. Will it make their daily lives easier? Help make training new staff simpler? Will it reduce customer complaints? Or generate more sales?

One discussion at the start of a project may not be enough to keep the commitment level high, especially when the project hits difficulties or is perceived to be taking a long time. Continually reinforce and update the message. Support from senior managers is a good way to demonstrate corporate commitment to the project. The team will be more committed to the project if they feel that their work is being recognized and supported from those higher up. Get the sponsor to drop an email to the team thanking them for their achievements so far and reinforcing the end objectives. Better still, ask the sponsor to mention your project in their next company briefing so the message that your project is important reaches a wider audience. Support from senior management is the second most critical factor in project success, according to a group of 236 project managers surveyed in 2002, after having clear goals and objectives.[50] If you can't reach into the upper echelons of the company, ask business users who are already starting to see benefits to attend a project meeting and explain the difference the project has made to them.

GOLDEN RULES

Try to create a culture of commitment and belief in the project's aims at the outset, and continually reinforce this throughout the lifecycle of the project.

26 Know the culture of your team

Every team has its own way of working, and when you start working with new people you need to appreciate the existing culture. Each team member will already have preconceived ideas about you and the others, at both an individual and a departmental level, before you even hold your first meeting. Trying to work out and understand these views will help you appreciate the culture of your new team and how best to make them work together successfully.

CULTURAL PROCESSES

Problems created by cultural diversity and the different ways in which teams work can be acutely obvious when your new team is based in another country, as Rasmus Kolind found out. He was sent to Spain for a year to set up a new process-improvement initiative in one of his company's call centres.

'We migrated a number of processes,' Kolind says, 'with the aim of improving them after the migration.' The Madrid call centre was handling work from four other countries, and Kolind had to implement the Six Sigma methodology for process improvement to increase efficiency. 'We were trying to implement something completely alien to the Spanish way of working, in a different language, and a different culture,' he says. Kolind, a Dane educated in the USA, travelled to Spain with two British colleagues and was very aware of the need to understand the differences in the way the cross-cultural team approached the project. They worked closely with the local Madrid team to lay the foundations for the launch of the new methodology. 'We had to take the approach "Let's understand what you do now" and not assume anything. They were a really close bunch of people, and initially it was important to be liked,' Kolind says. 'The first group we needed to convince was the Spanish leadership team.' He realized that it was essential to adapt to the local way of working in order to achieve anything. Kolind put his Spanish language skills to work, as he knew that conversing in the same language would be one of the secrets to success. This approach paid off. Kolind explains: 'The leadership team took time to get on board, but what was important was that we avoided saying "I know the answer . . . here you go."'

DEFINITION

Fons Trompenaars and Peter Woolliams define culture as 'a series of rules and methods which a society or organization has evolved to deal with the regular problems that face it . . . Culture is to the organization what personality is to the individual – a hidden yet unifying theme that provides meaning, direction

> and mobilization that can exert a decisive influence on the overall ability of the organization to deal with the challenges it faces'.[51]

It is precisely because culture is so ingrained in the way people work, and the way in which a company is organized, that it is sometimes difficult to spot what a team take for granted about how they do their jobs. Even a project team made up of people of just one nationality will have a particular culture, evolved from the organization's own corporate culture. Corporate culture manifests itself in many ways. Do they gel quickly and spend every Friday evening at the pub? If appropriate, consider going to the pub a few times so they get to know you. Do they keep information to themselves? If so, foster a culture of sharing information by encouraging them to outline their contributions to the project at your team meetings, and help them to see the advantages. Does the company have a culture of long tenure? Do they come from departments with strong hierarchical structures? If they have held their positions for a long time, be aware that they might resent you setting them tasks. Get used to making a quick call to their line managers so you can approach each new activity with 'I've run this phase past x and y and they agree the next steps are . . .' Even if you can identify what makes your newly formed team tick, how will this help you manage the project?[52] Being aware of the preferred style of your team will help you relate to them and will help you understand how to get the best from them.

HINT

Co-locating your team (having members based physically close to each other) contributes to project success. Research done by the US Civil Engineering Research Foundation shows that co-location contributes to effective decision making and attention to detail and helps the team form a partnership. Projects where the team members were not based together suffered from poor communication, procurement problems and lack of direction.[53]

GOLDEN RULES

Cultural awareness, whether inter-country or within a company, will help you get where you are going more smoothly.

27 Agree who is going to sign off

Who on your project will have the authority to make decisions at the most senior level? If you do not yet know, find out before the project gets too far downstream. You need to have someone to turn to in case you need a decision on the way the project is heading, and political power games can significantly delay (or stall or close down) a project completely. Thinking through the organizational structure of the project in advance can help prevent that.

a

TAKING THE LEAD FOR VETERANS

In 1998, the US Department of Defense (DOD) and the Department of Veteran Affairs (VA) started to work together on a project to share medical data for active-duty and veteran military personnel. The rationale behind this work was that as service personnel were highly mobile, they ended up with medical records at military facilities all around the world. An electronic system to share information between the two departments would help service personnel get the best care and also help process any disability claims.

The government computer-based Patient Record Project was started, but by April 2001 it was struggling.[54] The US Government Accountability Office (GAO) carried out a review and concluded that the departments involved needed to agree who was going to take the lead and be the main decision-making authority. This would establish a clear line of authority for the project, which would in turn make it easier to draw up and follow comprehensive plans.

By June 2002, the departments had still not agreed who was in charge, and the project was moving forward without coordinated plans, performance measures or clear goals and objectives.[55] The GAO once again recommended that a lead agency was designated to give the project a route for decision making.

Two years later, the GAO reported again that it was hard to establish what progress had been made with the two-way data-sharing project, as no lead entity had been identified, meaning that neither department had the right to make decisions that were binding on the other.[56] The initial aim of making military medical records easily accessible had taken six years so far and still seemed no closer to getting agreement or leadership with the way forward.

After reading the draft 2004 GAO report, the secretary of VA documented the department's response to the recommendations. He stated that the VA/DOD Health Executive Council would be the body that provided the final decision-making authority. However, by May 2005, no agreement had been reached regarding data sharing, leaving the VA uncertain that vocational rehabilitation services were being provided to all service members who needed them, and both departments still struggling to implement an

IT system seven years after they both agreed in principle that it was a good idea.[57]

HINT

It is essential to work out who on your project will have the authority to make decisions at the most senior level. You need to know who will be able to make the final call when you are asking for a decision on this or that solution, increases in budget, taking staff out of their day jobs for training and so on. Don't wait for an important issue to arise before you attempt to work out who in the organizational structure has the authority you require.

Break down the questions and you might find you need it to be several people: your sponsor may agree that staff training is necessary, but it could be the head of the operational department who decides when to make the relevant staff available. A survey by the Centre for Complexity and Change at the Open University found that a third of project managers are involved in the decision-making process but are not the sole decision makers. It also found that 24 per cent of respondents had no influence over the decision-making process at all.[58] If you fall into that category, it really is critical to work out who you need to turn to for decisions, so you can start to put together a stakeholder management plan and try to gain some input into the decision-making process, at least in terms of your recommendations being considered seriously. For more on stakeholder analysis, see Chapter 31.

The decision makers on your project should be authoritative and have the ability to negotiate with other key departments in case of conflict. Without a clear route for (binding) decisions, you could find your project stalling at the most critical moments, as the stakeholders disagree among themselves and refuse to abide by each others' decisions.

GOLDEN RULES

Work out who will be able to make the binding decisions in advance of needing to ask that person to do so.

28 Don't forget the soft stuff

Project management is largely about 'hard' skills: dates, money, resources, ticking off tasks. But to do those successfully, you need an appreciation of 'soft' skills: those elements of project management it is too easy to overlook when faced with time pressures and other crises.

a

THERE'S DELIVERING AND THEN THERE'S 'DELIVERING'

Patrick Mayfield worked for the UK government on the development of a best-practice method for managing projects, which was eventually launched as PRINCE2. Since then, he has been called in to help with many change projects that have struggled to get started. As chair of Pearce Mayfield, a programme and project management consultancy and training company, he has worked with many organizations and gives a typical example:

'A local authority made a policy to work more closely with the five NHS trusts in its area so that elderly clients returning from in-patient treatment were rehabilitated more quickly. The project leader broke down the policy implementation into a portfolio of projects to deliver refurbishment of three care centres, training of social workers, and an IT system to allocate and track specialist care equipment. However, the programme floundered and eventually failed to realize any of the strategic service performance objectives after an estimated three years and £4 million expenditure, despite the projects themselves being managed well and "delivering".'

What Mayfield found was that the programme leadership had approached the change as an aggregated project management exercise. It focused almost exclusively on tasks, deliverables and major end products. All those could be delivered without the programme being a success because key things were missing: an appreciation of the people involved, their interests in the programme, their motivation, habitual behaviours and culture, their power bases, and the necessary communications strategy.

'The paradox is that the "soft stuff" of change management is really the "hard" work of prevailing in change,' says Mayfield. Mayfield's example relates to a portfolio of projects, but the same task focus on a single project can lead to the same result. A project plan, a dedicated team and being on top of your risk register are no guarantees that your project will be successful. Your project will happen, but it may falter a few months later and, despite having delivered, be seen as unsuccessful. Mayfield has one further piece of advice: 'Make sure changes are embedded in operational areas: don't relax too quickly,' he says. He believes that this gives you the chance to ensure anything you have implemented has a fighting chance of still being there this time next year.

Soft skills are as much a part of the project manager's job as making sure tasks are delivered on time, and as much a part of the overall success as your end product. They include:

- stakeholder identification and management;
- understanding the culture of the team;
- putting together and using a communications plan;
- being present and available for your team;
- giving your team praise and guidance;
- getting buy-in from your team and also from the senior management – in practice, not just on paper;
- understanding who is accountable for decision making;
- ensuring people know why the project is happening;
- offering training to back up the communications if necessary;
- being positive and an agent for change yourself – never let anyone catch you saying the project is a waste of time!

Flick through the rest of this book for examples of these skills in practice.

Improving your soft skills will help you get the best out of your project team. Scholars have long been searching for the magic bullet that makes one project team more successful than another, and even within a single organization with a common culture and well-motivated staff some teams will just perform better than others. Marla Hacker has researched the theory that both the individual characteristics of team members and the dynamics within the team influence how the team will perform. She studied 22 teams made up of three or four people drawn from a group of university engineering students. The teams worked on a project over a semester and were graded in relation to the other teams, providing an incentive to perform well. To analyse whether individual characteristics made a difference to the team's overall performance, Hacker asked each student to complete a questionnaire about themselves and their experience. For the team dynamics part, the students rated their team's performance on 12 factors, including quality of discussion and level of agreement. Hacker's research concluded that only one factor made a difference to team performance – the students' average academic results – which she believes is representative of ability to accomplish tasks combined with the students' ability to motivate themselves.[59] She found no link between the students' demographic background, experience or team dynamics and team performance.

Project managers often wish they could handpick their team to get the 'best' performers. Hacker's research shows that it does not matter whether you choose your team or have them allocated to you because their individual social and employment backgrounds, and how well your group dynamics work, will make no difference to the team's ability to perform well. However, how far this research can really be applied in the workforce remains to be tested. Certainly an appreciation of soft skills and team interactions will make your job easier.

Soft skills are also important in a project environment because frequently the project manager will not have line-management authority for the people working on the project.

ANECDOTE

ITNET, a consulting and business process outsourcing company, started training project managers in 1995. Over the five years that followed, Peter Bishop, head of projects at ITNET, recognized that the company's training programme was missing out a critical skill for project managers: the ability to manage people as well as they could manage the technical aspects of running a project. ITNET brought in MaST, a training and development consultancy, to help redesign the project management course. A pilot course was delivered in 2001 and covered influencing skills, motivation, leadership, delegation and other soft skills side by side with the 'traditional' elements of the old project management course. The new course had a measurable effect on project managers and their projects: project managers reported being able to manage stakeholders more effectively. This had a knock-on impact for their projects. ITNET believes more projects are delivered on time and on budget and that customer satisfaction has increased as a result of the project manager's improved skills.[61]

GOLDEN RULES

Balance the technical aspects of project management with paying attention to the 'softer' elements to help embed the change and support the project team with the implementation.

29 Train your sponsor

You might be lucky enough to end up working with a sponsor who has already championed many varied projects, understands their role exactly and is keen to support this project and the team. Or you might find yourself working with someone who doesn't really understand what is required of them. If you find yourself in this latter situation, you need to take your sponsor under your wing and help them understand how their role supports the project team.

a

CUTTING THROUGH THE JARGON

In the 1990s, Neville Turbit began working with a major Australian government department to implement a project management methodology and tailor it over time to meet their needs. 'It was a big organization, where IT and the business did not enjoy the most harmonious of relationships,' he says. The department suffered from the same problems as many other large organizations at the time: prepare a project specification and throw it over the fence for IT to build. 'IT built it and – surprise, surprise! – it was not what was needed at the time of implementation,' Turbit adds wryly.

He knew that implementing a methodology alone was not going to solve the problem. 'I adopted a two-pronged attack on the business,' he says. 'First, we had to train sponsors to be involved and see themselves as decision makers. Second, we had to convince the rank-and-file business users that unless they were involved all the way through the process, they would never get a satisfactory outcome.'

Turbit, the principal of Project Perfect, a project management software, consulting and training organization based in Sydney, Australia, started with the management population who would soon become project sponsors. His 15 years of IT consultancy and a similar length of time in business roles has given him an insight into how sponsors react in project environments. 'In my experience,' he says, 'the biggest problem is getting the person who is responsible for the project to understand what their role is all about.' Turbit feels that sponsors are out of their normal operating environment when they are working on a project, and that can block them from being effective. 'Many sponsors have little idea why a project is different to any other department they might be responsible for. I have had one sponsor say to me: "I don't know what you want from me. You are more demanding than any of my line managers." I had to point out that I was doing a non-routine job. His line managers had clearly defined processes and procedures and they just tweaked the business process. I was creating it.'

One sponsor confessed to Turbit that she had no idea of what she should be doing. 'She felt so uncomfortable she basically delegated all the decisions to IT,' Turbit says. 'When I first talked to her about using a methodology, it was not much better. Once again, it was a language that didn't sink in.' Turbit continued to work with the sponsor, helping her understand her role and the 'world of projects'. 'The breakthrough came when I took her a decision on a project,' he recalls. 'I told her that she needed to make the decision. I said: "It is your responsibility. Our responsibility is to present you with options and not leave here until we explain the problem and the options to resolve it in language you can understand." Evidently we spent considerable time discussing the issue in layman's terms.' Turbit stopped talking in project management jargon, gave up using the words 'databases' and 'functionality' and started talking about what the new system would do, and what things the team needed to keep track of. 'She said this was the first time she did not feel foolish in front of her subordinates discussing the new IT system,' Turbit says.

Turbit made sure the rollout of the project management methodology at the government department included time for sponsor training. The focus was on getting project managers to talk the language of the business 'rather than techno-babble'. He also made sure project managers knew never to take a problem to a sponsor without a number of solutions. 'Over time, we built up a core of people who sponsored projects. They knew the questions to ask and were not afraid to become involved in providing solutions,' he says. 'It's taken a decade, but this particular department now has the most involved business resources in any organization I have seen. Project scopes are developed in an orderly manner, and budgets are realistic. It is always interesting when new people come into the organization. Typically they have a better way of running projects. The first person to tell them to "do it our way" is typically a business person. The business people have seen the approach work, and the sponsors have sponsored projects that they truly controlled.'

DEFINITION

Eddie Obeng in his book *Perfect Projects* defines the sponsor as 'a person who:

- invented the idea and really wants to do it;
- controls the money;
- wants the end product or will end up living with it;
- can provide effective high-level representation, and smooth out the political battles before you get to them;
- "owns" the resources;
- acts as an effective sounding board/mentor.'[62]

Unfortunately, you might not be able to choose someone who meets all these criteria. People in the role of sponsor also suffer from having many different demands on their time, and your project may not be on top of their list. If they don't know what it is they have to do, you can be sure they won't be able to make the time to find out. You have to be there to help them discover what being a sponsor means and to explain what you expect from them.

Some organizations use sponsor training and expect all their senior executives to attend. If formal training is not available, or if it has been a while since your sponsor attended something like that, it falls to you to offer some guidance. The beginning of the project is normally the best time to introduce a sponsor to their role. Try to find out what experience they have had, what went well and what they found difficult about sponsoring previous projects, but be warned that senior managers may not be willing to share their experiences with you. You can still approach the subject tactfully: 'I know you've already sponsored loads of projects, but as we haven't worked together before I just wanted to explain to you how I see your role as the project sponsor, and then we can establish how best we can work together.' Use the sponsor FAQ (frequently asked questions) below or make up your own as a starting point for discussion.

SPONSOR FAQ

Q. Why does a project need a sponsor?

A. To support the project team and act as an escalation route for any issues or problems.

Q. How is a sponsor different from a project manager?

A. The project manager manages the operational, day-to-day issues on the project. When something happens that they can't manage within the agreed parameters (a budget, a timeframe, a set of requirements), a sponsor makes the decision about how to proceed.

Q. How can they make the decision when they don't work daily with the detail?

A. The project manager will present various options and the consequences of following each option. The sponsor should have a general overview of the project, which will be enough to choose the right course of action.

Q. How is a sponsor informed of progress?

A. This is agreed between the project manager and the sponsor at the beginning of the project. It could be a written monthly report, a face-to-face briefing or on an exception basis. If there is a problem that cannot wait, the project manager should be able to approach the sponsor immediately.

Q. What else does a sponsor do?

A. A sponsor:

- represents the project at a senior management level;
- keeps the project manager informed of any changes or developments that may have an impact on the project;

- puts their name to and helps with communications about the project;
- offers advice and makes decisions;
- puts forward and/or supports the case for a comprehensive budget for resources;
- chairs the steering group;
- reads, understands and signs off project documents;
- does anything else (within reason!) to support the project at the request of the project manager.

A sponsor may also need 'training' in the more technical elements of the project. As the work continues, you will soon become an expert in the intricacies of what it is your project is delivering, but you cannot expect the sponsor to understand the details or the jargon. Present your project updates with clarity, keep the use of jargon to a minimum, ask open questions to test their understanding and give them the opportunity to ask you questions too.

GOLDEN RULES

Don't automatically assume your sponsor knows how to carry out their role effectively. Find out and explain what you expect from them.

30 Bribe your team

We're not talking about dabbling in the realms of the illegal here. Sometimes teams just take a little bit of encouragement to get started or as an incentive to up their game at critical points in the project.

a

MASSAGING THE RESULTS

Starsys Research Corporation provides components to the aerospace industry. It starts and completes approximately 40 individual programmes per year, ranging from $10,000 actuator builds to several-million-dollar mechanism programmes. The average project duration is about nine months, so the company has an opportunity to try out, observe and iterate different approaches to project management.

One of their approaches was to track informal agreements between staff – the kind that got mentioned in passing during a project team meeting and normally escaped the minutes. Once the mechanism for tracking these was in place, it was soon clear that only half of the 'I'll send you a copy of that spreadsheet this afternoon' arrangements were ever actually honoured. President of the Boulder, Colorado, company, Scott Tibbitts, decided to combat this by incentivizing his staff: he told everyone that if the results stabilized at 75 per cent or better for two weeks, he would bring in two masseuses for a day. The next day, the tracking mechanism showed that 77 per cent of informal agreements had been met. The numbers stayed up for the following two weeks and, true to his word, Tibbitts hired the masseuses as a reward for the increase in productivity.[63]

Bringing in masseuses may be a little out of your project's price range, but you can use the same principle. The basic premise is to offer your team a reward in exchange for their efforts on some of the more boring project tasks. And there will be boring tasks. Projects thrive on creativity, dynamism and motivation, but unfortunately there are also bits that seem to go on forever without resulting in any progress.

You do not have to invest heavily or plan anything too elaborate to offer a small incentive to the project team. Introducing food to any project team situation is normally an easy win:

- Provide biscuits or a box of chocolates for long workshops. Bring them out at half-time when you need to energize the group and provide a mental break.
- Allow some budget provision for occasional treats. A four-hour planning session in a hotel meeting room with no money for tea or coffee is destined to be a struggle.

- If any groups have to work weekends or unsociable hours, go in on occasions yourself and offer to do a pizza run.
- Hold short dull meetings with small groups in a coffee shop (or even the staff canteen). In an off-site environment, be careful to respect business confidentiality: do not shout your company's strategy across the room or leave papers in view.

GOLDEN RULES

Use occasional treats to encourage your team through the tough or boring parts of the project.

31 Carry out stakeholder analysis

Your project will touch the working lives of many staff. You will quickly identify some of the groups who will be affected, but there are probably others who have an equal influence over the success of the project. Stakeholder analysis is a tool that helps identify all the relevant parties and map their interests and influences over the project.

a

IT'S ALL ABOUT POLITICS . . .

Early in 2005, the management board of a top Dutch company decided to take action to boost their performance in the marketplace. They set up a large programme of activity aimed at increasing the focus on service in the company.

'Although you might think that the department managers were aware of the problems, the opposite was true,' says Paul van Doorne, a programme manager with 12 years of experience in project and operations management. 'The managers didn't accept the fact that we were one of the worst-performing companies.'

At the outset of the programme, the team concentrated their efforts on getting the majority of their stakeholders – the management population – behind the idea of improving performance with large-scale presentations. They also collected data about how long processes took within the company and measured how many errors were made.

'The team pinpointed the problems,' van Doorne says. 'There was no interaction between the departments and scattered responsibility where nobody was responsible for an end-to-end process. Targets were not set by what customers wanted and thought were acceptable but were focused on meeting internal administrative goals, and management information, where it was available, represented only the work of individual departments and didn't reflect the end-to-end customer experience.'

Once they had their evidence, the team were able to present it to the management population. The response was not quite what they had anticipated. 'One might expect that the alarm bells would ring and every manager would be supporting the programme,' says van Doorne. 'In fact, the opposite was again true. The managers started to question the data we had analysed and argued that it was not **their** part of the process that was not performing.' To add to the internal politics, the management team were united in one thing: the problems were the fault of a part of the business that had been recently outsourced, and this was now the common enemy.

Van Doorne continues: 'After six months pushing the managers to undertake actions, the actual improvement in company performance was close to zero. Like most change management situations, the majority of the effort went into influencing the staff who had come down against the change and not on the 20 per cent who were already supporting the change.'

The team decided to take another approach. They conducted a stakeholder analysis, which showed them the current level of support for the performance improvement initiatives and their willingness to change for each individual or group. They also developed a strategy to get each of the stakeholders on board. From their analysis, the team identified one manager who was willing to implement changes to improve performance in her team but was not sure how to go about it. She was new to the operations department and had a sales background.

'We helped her set up a basic process measurement system,' says van Doorne. 'Her department made progress very quickly once they had a way to measure how they were doing and identify areas of weakness, and once she saw the difference she was very supportive of the whole programme.' The team called on her to discuss the improvements in her team with other parts of the organization. 'Her sales background meant she was excellent in getting the message across,' he adds.

The programme team identified another department that seemed to be pivotal for every customer-centric process, which was headed up by a specialist. 'He was brilliant in his field, but he was not a good process manager,' says van Doorne. 'We decided that the processes in his area needed to be industrialized.' Having a clear view of how the department operated and the key players within allowed the programme team to focus on influencing the relevant people. 'The programme manager started to influence the chief operating officer and the board, the project manager worked on the departmental manager. Other team members who were active in departments that were touched by the process also had a chance to express their views,' van Doorne says. 'As a company, we recently took the decision to appoint a process manager next to our specialist. Now the process can be run like a factory, which is better for the customers as we can track our performance and we know how to improve upon it.'

Van Doorne is very clear about how stakeholder analysis helped the improvement initiatives. 'Stakeholder analysis is a very powerful tool,' he says. 'The most important thing is that the analysis makes you aware of politics and positions and therefore you can influence people to get to your end goal.'

Stakeholder analysis or stakeholder mapping is usually done in three stages:

(i) Identify stakeholders.
(ii) Record their position (interest) regarding the project.
(iii) Compile an action plan for how best to influence them.

DEFINITION

Boddy and Paton in their 2004 research define stakeholders as 'individuals, groups or institutions with an interest in the project, and who can affect the outcome'.[64]

It will be straightforward to identify the inner circle of stakeholders: you, the team, the sponsor, the project's customers. You do need to go beyond that though, so draw on your team and use their ideas as well to make sure that all relevant parties make it on to the list. For example, stakeholders can also be external groups such as government bodies, regulatory agencies or people from other commercial organizations.

The next step is to plot their current interest in the project using a chart such as Figure 31.1. Don't ask stakeholders directly where they see themselves: they will probably give you a politically correct answer. Base your evaluation on what you see them do and say. Then work out where you want each individual or group to be in terms of their support for the project. You will never need all your stakeholders to be 100 per cent supportive all of the time. If your chart reflects that, you have probably missed out some of the smaller stakeholder groups.

Identifying stakeholder groups is the easiest part of the analysis.

At this stage, consider the relative power and influence of each stakeholder. If all your critical high power stakeholders are currently not showing much interest, you have already identified a risky situation that needs addressing quickly. What can you do to address it? Communicate, attend meetings, explain the benefits, find other people who can sell the project on your behalf, find allies, network and use whatever influencing techniques work for you.

'The ability to understand the often hidden power and influence of various stakeholders is a critical skill for successful project managers,' write Linda Bourne and Derek Walker in their research paper about stakeholder mapping. 'Stakeholders can be a considerable asset, contributing knowledge, insight and support in shaping a project brief as well as supporting its execution.'[65] To make the most of your stakeholders, move to stage (iii) in the list on page 110 (compile an action plan for how best to influence them) to identify what as a team you can do to move each stakeholder from their current 'interest' position to where you need them to be. Figure 31.1 shows that the seven sales support staff have the furthest to go in terms of changing their attitudes, but they are a group of low-power stakeholders. In this situation, the project manager would need to tackle this group but would have to balance the effort required with ensuring the IT manager, a high-power stakeholder, is also influenced, as this role appears more critical to the project's success.

Stakeholder	Averse	Uncooperative	Indifferent	Supportive	Enthusiastic
Sponsor				□ ⟶	◆
Finance director	□ - - -	- - - -	◆		
IT manager		□ ⟶	⟶	◆	
Database analyst			□ ······	◆	
Training team (3 people)			□ ······	◆	
HR supervisor		□ - - -	- - - -	- - - -	◆
Sales support staff (7 people)	□ ······	······	······	◆	

□ Current position
◆ Desired position
⟶ Highly influential stakeholder
- - -▸ Influential stakeholder
······▸ Uninfluential stakeholder

FIGURE 31.1 *Mapping stakeholder interest and influence*

You can include a lack of interest from stakeholders as a risk on your log if you want to manage your action plans formally. Track the actions required to influence each stakeholder as you would do for any risk-management plan. Just be aware that the risk log is not a private document and it may be more astute to keep some of those activities within the immediate team. Stakeholder analysis and the subsequent activities can be political and organizationally sensitive.

SIX STEPS FOR ONGOING STAKEHOLDER MANAGEMENT

(i) **Identify stakeholder groups.** You can't start managing stakeholders until you know who they are. Think of the main groups or departments affected by your change project. Launching a product will of course touch marketing and sales, but what about training? Opening an office will affect HR and IT, but what about the switchboard operators and the internal mailroom? Carry out stakeholder analysis to understand their interest and influence over the project.

(ii) **Nominate a key contact**. From each of those groups, pick someone to be the key individual. Be careful who you choose. What are the internal politics of picking one over another? Your key contacts should ideally be people who are affected directly, with enough authority to make decisions about things that touch their departments.

(iii) **Feel what's going on**. Contact your list of named individuals and introduce them to the project properly, dispelling any myths they might have held. Get an accurate understanding of their position regarding the project, and validate the assumptions you made about their interest and influence in step (i).

(iv) **Observe them.** The more you work with and involve your stakeholders, the more chance you have to observe them. Don't let stakeholder groups drift away. As the project progresses, act to bring them further into the project in a positive way, observing how their attitudes change and adapting your action plan accordingly. Your aim is to work out each negative touch point and address it, moving the stakeholder's negative attitude to a positive one.

(v) **Review what's going on**. People and job roles change. So do projects. The person who was suitable to represent the legal team 10 months ago may not be the right person today. Don't be afraid to ask your stakeholder whether they still feel relevant to the project – and if they are not, ask who should take over from them. Brief the new addition to your project team on their role and responsibilities, decisions in the pipeline and what decisions you will expect of them in future.

(vi) **Manage the process.** Finally, monitor and manage your stakeholders and their expectations as the project progresses – not just at the outset and when you need something from them. A quick call every now and then (put it in your diary) just to keep them up to date will go a long way to promoting goodwill and building your reputation as an excellent project manager. At the end of the project, thank them and manage them out of the team. Managing someone out of the team means they understand that the project organization is coming to a close. Each ex-team member or stakeholder should understand the process for raising future issues or questions through their new business-as-usual contacts.

The key message: the acronym formed by these six steps is INFORM. Ask your stakeholders at the outset how they would like to stay informed, and make sure you do it. Do it even if they say they are not bothered about receiving information about the project. One day they will be interested, and then they will know exactly how to reach you for more information.

GOLDEN RULES

Mapping your stakeholders' positions with regard to interest and influence will:

- quickly highlight any immediate concerns;
- allow you to draw up an action plan to address those concerns.

32 Present your stuff interestingly

As a project manager, you will have to explain the details of your project to people who are not project managers or project team members and people who have very little time to listen to what you have to say. By thinking about what you say and how you say it, whether it is for a short update meeting or an hour-long presentation, you can engage your audience more effectively.

a

MAKING YOUR MARK

'Our director had a project day,' Catalina Marcos says. 'He wanted the project managers to each present what they were working on. By the time my bit came round, he'd already sat through 15 short presentations.' Marcos had predicted her director would be tired of looking through printouts of PowerPoint presentations. She decided to put her project updates across differently. So she turned to Microsoft Word and the huge range of standard templates available online. For one project, she created a three-fold brochure. 'I only had to design the cover. The rest of the text I copied and pasted from the project brief and other documents,' she explains. The other project Marcos was working on had been in difficulty for some time. 'I summarized the current position, risks and outstanding decisions at the top of a landscape sheet of paper, and underneath created a timeline to visually show the project history, including all the false starts and delays,' she says. 'Setting it out like that was much more effective than several paragraphs explaining why we were where we were, and also highlighted the decisions I needed him to take and the reasons I needed an answer quickly.'

Marcos believes that presenting the information like this took no more time than pulling together a presentation on slides. 'I just thought about what I would want if it had been me sitting through all those updates. It gave our director a break from the same format, and hopefully a reason to remember me and my projects.'

Whether you are presenting information to your team or senior management, bear in mind your audience and try to make your work interesting and memorable. You probably have to complete standard monthly progress reports, and changing the format of those could cause problems for the people who receive and use the information. However, you are at liberty to be creative with ad hoc reports and presentations. Some ideas:

- Has a task been outstanding on your meeting minutes too long? Use your word-processing package to make it flash when the document is viewed electronically. This type of humour can speed up results but can also backfire, so consider the person on the receiving end of the flashing task and be sensitive to how they will react.

- Doing a stand-up presentation with slides? Make it professional, using your company or team's standard presentation layout. Not got one? Set up a template. Use slide builds and interesting transitions between slides. Incorporate video clips, screenshots or a live system demonstration, if it would be relevant to the content of your presentation. Get someone to proofread it: your audience should be captivated by your fascinating presentation, not on the lookout for the next spelling mistake.
- If you have to send out your presentation in advance, and then intend to talk through it at a meeting, be sure to have some extra information to give. Reading out what your team have already seen is certain to create a few droopy eyelids.
- Sending a status report to someone who knows nothing about the project? Check it for jargon but also tone. Replace passive sentences with active ones: 'The system testing was carried out by the German IT team' becomes 'Our German IT colleagues carried out the system testing.'
- Set yourself challenges. Can you halve the number of slides? Can you shorten the report by two pages? If you print the update on both sides of the paper and make a booklet, can you get it all on one sheet? Keep it short and simple: include the information the recipient needs to know, not a list of everything the team has done since the last update.

Be sensible and remain professional: no holiday pictures, inappropriate informality or submitting your child's playgroup drawings as a visual expression of project progress. Avoid peppering any document with clipart-type pictures.

GOLDEN RULES

Consider your audience when giving project updates, and brush up your presentation skills: see the further reading section for some ideas of where to go next.

33 Organize a party

One of your roles as a project manager is to keep the team motivated throughout the project, so they continue to meet deadlines and turn in quality work. The end of a project, after your final team meeting, can seem like an anti-climax to those people who have been working on it for months, sometimes years. They will quickly forget any little treats or motivational rewards you organized during the project. You need to do something else to mark the end.

It is important to give the end of your project a sense of finality. This is especially important if your project got canned. Team members are likely to feel that they have wasted their time and that their efforts went unnoticed. The official end of a project, whether it ran its natural course or was terminated early, is a good time to recognize everyone's commitment and achievement and to thank them properly for their work – what better way to do that than have an 'It's-a-wrap' party? Movie makers have been doing it for years.

ANECDOTE

Sam, a project manager in a large electrical engineering firm, had spent 18 months leading a 14-strong team of business and technical experts in delivering a new database system. The database allowed clients to look up the status of their orders online and within a few weeks of launch was already making a difference to the number of queries received by the customer service teams. The project got a small write-up in the internal magazine, but Sam knew that her close-knit, dedicated team – six of whom had given up many Saturdays and evenings testing the database – did not really feel that it was over. They needed to close the piece of work mentally as well as on paper, so Sam approach the company's HR department for advice. She found out that the firm had a central budget for rewarding high-performing teams and got an application form. Supported by her project sponsor, Sam completed and submitted the form. Two weeks later, she had authority to spend £30 per head on an event for her team, with the bill going to HR. She negotiated a deal with a local restaurant and the 15 of them went out for dinner one evening after work, as a fitting end to a successful implementation.

A formal recognition scheme, where such schemes exist, is a great way to offer your team a celebration for a job well done and also to provide a firm feeling of closure. However, even if your company does have a recognition scheme, you might not be successful in applying for funds. For example, your project may have been closed down prematurely or the funds may be very limited and you might be unlucky. However, you can still achieve the same result without spending a lot:

- Ask your project sponsor whether they can provide a few bottles of wine or soft drinks for an after-work reception. The sponsor could also say a few words of thanks.
- Organize a pot-luck picnic, where each team member contributes an item of food or drink, and take a long lunch break in a local park.
- Does anyone in the team have useful contacts? Somebody with a large garden may be prepared to host a barbecue.
- Ask whether your local wine shop could arrange a tasting.
- What awards are available to enter? It could take several months for the judging to be complete and of course you could lose, so you need to do something else in the meantime. However, awards are great for the CV and promote a real sense of achievement. The trade press will have details of annual awards relevant to your industry. Keep an eye on the local press too for news about local competitions.
- If all else fails, get the team together for a group photo and circulate copies to everyone with a personal 'thank you' note from you, summarizing the main successes. The notes will mean more if you tailor them specifically to the addressee instead of sending out a circular memo. If you are pressed for time or words, a printed memo to everyone with a handwritten 'Thanks for all your hard work' on the bottom will just about do it.

ANECDOTE

However you decide to thank your team, make sure everyone can participate. A night down the pub may not be the right event for a multicultural team. One project manager organized a (rather ambitious) tank-driving event before realizing that his disabled colleague, who had been a key player in the project's delivery, would not be able to take part.

GOLDEN RULES

Closure of the project is important, but so is closure of the team as a unit. Mark the delivery or closure of the project with a notable moment to both provide a sense of 'ending' and to celebrate your work together, even if the project did not make it to the final implementation.

34 Find out what motivates your stakeholders

DEFINITION

Stakeholders are those people who have a vested interest in your project, whether they want it to succeed or fail. Winning over the people who fall into the latter category will really make it easier to run the project. One of the ways to do that is to establish what they want from the project, and find a way to give it to them.

STAKING IT ALL (PART 1)

When Lotterywest, Western Australia's state lottery, embarked on a large rebranding exercise, project manager and head of the shop-fit team Keith O'Shea knew that the stakeholders would be key to the success of the project. His main group of stakeholders were the retailers who would be adopting the new aqua-and-yellow brand. To keep them engaged for the duration, O'Shea and his project team looked at what would interest those involved and realized that they wanted a few straightforward things: help to cover the costs of the rebranding, help with marketing and to be informed regularly of what was going on.

Lotterywest responded to these needs by putting several strategies in place. The company organized interest-free loans, which were made available to all the 484 lottery outlets. Lotterywest and their advertising agency also ensured that all television and newspaper adverts featured the new brand, even before the first shop-fit had been completed. This enforced the new image in the public domain and helped convince retailers to move towards it. 'We held a public briefing for industry, people like designers and shopfitters, to engage them in the "selling process",' O'Shea adds. 'This proved to be amazingly successful, as they became the main drivers in the initial uptake. It also made the retailers aware that the changes were happening for real and happening now.'

He took it upon himself to ensure he was in regular contact with all stakeholders. 'I phoned them all regularly,' says O'Shea. 'The team and I visited the outlets in person, explaining how easy it would be to comply with the new branding and doing some communication management – dispelling any myths that were in circulation,' he explains. 'We had a really good response to this approach as the shop-fits were being completed at the rate of four per week and the newly branded shops were reporting extra sales.'

Find out what your stakeholders want. What would motivate them to do their part in the project? Recognition within the company? Their staff to be paid overtime? Do your research subtly, as asking someone outright what they would want to guarantee their participation could come across the wrong way. Instead, try to think about it from their point of view. Your stakeholders will have day jobs and targets, and your role is to make this change as painless and straightforward for them as possible.

HINT

A person can be motivated by several things. It may be easier for you to concentrate on one thing, such as personal ongoing contact, and deliver that well, instead of trying to manage all their needs poorly.

Once you have worked out what you believe motivates your stakeholders, find out how your project can give it to them. It could be something outside the project, such as less time spent by the department on reporting. Can your project deliver that as a benefit? If not, what is the closest you can get to it? Never lie about what your project can deliver for them, as at the end they will find out and be both disappointed and cross. But do consider how you could rephrase some of your project's stated objectives to better reflect your stakeholders' needs.

GOLDEN RULES

Your stakeholders are critical to the success of your project. Find out what they are motivated by and, if possible, give it to them in return for their participation.

Section 4
Managing project plans

Moreover the hastening of any matter breeds disasters, whence great losses are wont to be produced; but in waiting there are many good things contained, as to which, if they do not appear to be good at first, yet one will find them to be so in course of time.

Herodotus (484–c.425 BC), *The Histories, Vol. II*

Project planning is the process of identifying what tasks need to be done in order to complete the project and to meet the project's aims and objectives. Planning gives you, as the project manager, the opportunity to ratify with the stakeholders that you have really understood what they want and that they understand what they are going to get.

Scheduling is the process of putting these tasks into the correct order. Schedules are flexible, and it is up to you how you define 'correct'. Probably what is 'correct' when you start the project will not stay that way for long, and you will end up modifying your schedule as you go through the project.

Most projects are less than a year in duration, and only 14 per cent of projects take longer than 18 months.[66] Even short projects need accurate plans and a managed schedule, otherwise you will find them taking a lot longer than was ever anticipated.

This section deals with planning, scheduling and time management in general, which are all key skills for project managers.

35 Keep up the momentum

Starting up a project is often the easiest part. Keeping it going takes a lot of effort. Putting some thought into how you will keep the project's momentum while it is still in the honeymoon period will be time well spent.

a

STAKING IT ALL (PART 2)

Keith O'Shea joined Lotterywest, Western Australia's state-owned lottery, in September 2003 as head of a new team with the task of updating the look of the 484 lottery-ticket sales points across Australia. He had to hit the ground running, as the first retail makeover was five months later, in February 2004. 'After the first one, it was clear that we needed to maintain the momentum,' says O'Shea, an Australian project manager with 20 years' experience in various industries, including construction, manufacturing and computing. 'Other lottery jurisdictions had embarked on similar campaigns and started well, only to stall after a short period of time and a proportionately small number of retail outlets made over.'

Not wanting to make the same mistakes as the others, O'Shea and his team used various strategies to keep the rebranding project on track to ensure it did not come to an abrupt halt before its planned completion in February 2007. 'We had to be especially thoughtful in how we motivated retailers to move to the new brand,' O'Shea explains. 'It was the outlets themselves who had to find the money to pay for the change.' To help the retailers along, the project team produced a 'makeover kit' that was made available to the rural and remote shops. O'Shea travelled the country running workshops to engage the store mangers and staff with the new image. The project team also organized a celebration for the 100th store to take up the new brand, and the ensuing party was covered in the internal magazine, which dedicated a page of each bimonthly edition to news about the project. None of these things was a huge overhead for O'Shea's team, but all were essential in keeping the project moving along. 'I have likened the possible stalling of our project to the stalling of a jumbo jet – very difficult to kick-start,' concludes O'Shea. 'We knew we had to keep up the pace, first to get it all done, but second to carry the staff along with us.'

Relying on others is one of the main reasons projects falter. Other people have day jobs and priorities that do not necessarily align with those of the project. Consequently, a good project manager will be able to get them to do their part of the project without it being at the expense of their day job.

The trick is to make it easy for those outside your direct control to do their part, while also making it harder for them not to get involved. For example, if your team cannot follow the Microsoft Project plan you have produced, transfer it to a bullet-pointed list. Ask them to tell you when they will complete their tasks: if you impose a date, it is easy for them to say it was unachievable. It is much harder for someone to explain why it has not been possible to meet a self-imposed deadline. Communicate what others are doing and the benefits they are receiving. Offer as much help as you can while subtly increasing the peer pressure and finding answers to any excuses you hear as to why things are not progressing.

There is one big risk in situations where the momentum of the project is slacking because people are not doing what is required of them. Letting things slip over a prolonged period results in a stalled project that could be impossible to start up again (and the blame for its premature closure on your career record). 'Get it done fast,' write Robin Lissak and George Bailey in their book, *A Thousand Tribes*, about their experiences at Pricewaterhouse-Coopers:[67] 'In a large organization, staying the course on any firmwide initiative requires speed – 30-, 60-, 90-day outputs – or it rarely reaches fruition. Unless the game plan is based on speed, a company tends to add time, effort, and bureaucracy to a project so that it never gets done.' If you notice things starting to slow down, flag the deceleration to your sponsor as an issue.

GOLDEN RULES

Be aware that any slowing down of activity could be the first sign of project demise, so help your team to keep the pace.

36 Plan first – set end date later

A project is always someone's idea, and that person always has an idea of when they want it completed by. One of the hardest things to do as a project manager is to manage the expectations of your sponsor, key stakeholders and team, who will all want to know when this project is going to finish before you are ready to tell them. A period of detailed planning at the beginning of the project is essential for two reasons:

- it provides clarity about what it is that you want to actually achieve;
- it gives you a firm foundation and confidence in your schedule dates.

After your planning activity is complete, you can announce, with as few caveats as your risk management allows, the amount of time the project will take and therefore when it will be finished.

PLANNING FOR PUBLICATION

It's difficult to predict how long the publishing process will take when you don't know what or how many submissions you will receive. That's the dilemma faced by the editorial board of the Parisian literary magazine *Upstairs at Duroc*. The planning process is dictated by the volume of work, which is known only after the submission date passes. The magazine receives over 400 short stories and poems each year and, once submissions are closed, the planning can begin in earnest. 'We can't confirm the publication date until we really have an idea of what selection process we have to go through,' explains Barbara Beck, editorial director.

Two teams of dedicated volunteers read poetry and prose submissions and sift out those that do not make the grade. This process can take several months, depending on the volume. Then there is a second reading, where the editorial board reviews all the remaining pieces over a number of weeks, makes detailed notes and then meets to agree the final selection. The prose team normally finishes first, as the magazine receives fewer stories and the board often has very similar ideas about what makes a good piece of narrative. The poetry decisions are more challenging. Arranging the final editorial meeting is difficult, as it needs to fit around the availability of board members, which can lead to more delays.

Once this is done, the chosen pieces are handed to the layout team, who prepare the magazine's 100 pages for the printer. A talented artist produces the front cover, and the magazine is set to go. The publication date can now be agreed, and the printer can confirm a date for delivery of the finished magazines. 'We can use last year's schedule as an idea of how long things will take,' Beck says, 'but we can't assume it will be exactly the same. We may need to commission new fiction or poetry if we don't receive enough quality submissions. It's unconventional if you compare it with the way monthly magazines are produced, but we are an unconventional magazine and we don't work in the same way as a high-street glossy.'

To arrive at a position where you understand the overall duration of the project, start with the basics. Sit down with your team and work out the project plan, applying the same rigour and techniques to project planning as you do to the rest of your project activities. Take into account the advice in this book and best practice in your own organization. What needs to happen? How long will each task take? The answers to these questions will allow you to pull together a timeline.

HINT

In all but the most exceptional of circumstances, try to use estimates gleaned from the people who will actually be doing the work. If you cannot get hold of them, or someone with a similar profile or background to them, make an educated, conservative guess. Educate yourself if need be: find out whether a similar project has been done before. Can you speak to the manager? How long did that piece of work take to complete?

It is important to be realistic about what tasks need to be done and the length of time each task will take. In a 2002 survey, having a realistic schedule was identified by 78 per cent of project managers as critical to a project's outcome.[68]

HINT

It is a fine balance between spending too long at this stage and meeting the expectations of your stakeholders: you cannot keep them waiting forever. Aim for a best estimate end date, and tell them that's all it is until you have a chance to plan in more detail. The more accurate you can be now, the easier it will be later, as the finish date should not move too much.

It is always easier to plan the early stages than the later ones. You know what you need to do now in order to get things moving in the short term, but predicting what activities your team will need to do six months down the line or even later (with all the changes and modifications that come with managing a project) is much harder, to the point of sometimes being impossible. It is still possible to plan at a high level. Breaking down your project into phases or stages will help. Plan the first phase in detail and have a high-level plan with perhaps just a few important milestones for the subsequent phases. Phase 1 should include the planning tasks for phase 2 and so on. Your high-level plan will have enough details to satisfy a sponsor and give an overall impression of when the project will deliver, but be qualified by the need to do more detailed planning later. As the plan for each subsequent phase should be signed off by your sponsor, they will be obliged to agree and understand the flexibility within the end date.

HINT

There will be occasions when someone on the way to a senior management meeting stops you in the corridor and asks you to provide a date on the spot, before your planning is complete. If you truly cannot get out of answering, be as vague as possible. Say 'By the autumn' or 'During quarter one next year'. If pressed, use your best guess as of that moment, plus some extra time as a safety margin. It is nearly always easier for stakeholders to manage the communications around a project that delivers early than for them to explain why it is late. The company's management team will not know (and will not want to know) the details of your project, so they have to take your word on the duration.

GOLDEN RULES

Don't get sucked into the trap of promising an end date before you have really worked out the tasks involved.

37 Manage fixed-date projects carefully

WARNING

Some projects are already time-bound when you receive them. Maybe there are regulatory requirements to meet by a certain date. Perhaps your chief executive officer (CEO) has promised a major customer something by the end of the year while they were out on a corporate golf day. Projects with fixed end dates present a different type of planning challenge for project managers. Instead of being able to analyse and plan, you are told what to do and when to do it by. Some degree of direction is good – after all, you cannot justify a three-month planning phase for a project that ends up taking only six weeks. The analysis part of planning has to take on a different spin when your implementation time is already ticking away.

SAVE THE DATE

Stephanie Haefner, a floral designer who runs Bridal Blossoms in New York, is used to having to plan her activities around a fixed date. The bride and groom arrange their wedding date to suit themselves and their families, and she then has to react and plan accordingly.

'Some brides call me a year in advance, but most call between six and eight months prior to the wedding,' she says. This normally gives her plenty of time to plan, as many of the tasks have to be done at the last minute to keep the flowers fresh. 'When the client leaves, I take the handwritten copy of their wedding flower order and transfer it into my computer,' Haefner explains. 'I don't have a program for it – I haven't found one I like. I use WordPerfect and type everything up myself, just the way I like it. I list at the top all their personal information, address, phone number, locations of ceremony and reception and the times they are starting.'

Planning backwards from the wedding date, Haefner establishes the order for the other post-sale tasks. Ordering the flowers is done a little in advance, with delivery three days before the wedding. 'It gives me time to let the flowers hydrate, and if there is a particular variety, such as a lily, that needs time to open to the perfect stage, I need a few days to allow for it,' she says. It also provides a contingency period in case of a problem with the delivery.

'Two days before the wedding, I'll start preparation,' she continues. 'I make bows for corsages, get ribbons ready for bouquets, and pretty much do any prep work I can that does not involve the actual fresh flowers.' The flowers are arranged the following day and packaged ready for delivery.

This is a plan that can be repeated for the majority of weddings, and Haefner appreciates that working to a fixed date gives her the opportunity to focus and reuse a schedule that she has already proven many times. 'I take every wedding one at a time. I hate to do too much ahead for fear of getting weddings mixed up,' she says.

But occasionally, things happen without so much notice. Haefner was planning the flowers for a small second wedding booked with just six weeks to go. 'I ordered the flowers well in advance, three weeks before the wedding, even without payment,' she explains. 'With the wedding planned for 30 December, in the middle of the holidays, I wanted to make sure I could get the lilies the bride wanted.'

The week of the wedding arrived and Haefner had not had confirmation from the bride or received payment. 'I didn't want to assume she didn't want the flowers,' Haefner says. After many phone messages, the bride finally called back. 'She had decided the lilies were too much at around $250 and asked if I could do something less expensive.' Three days before the wedding, Haefner cancelled her original delivery and placed an order for some cheaper roses. The flowers arrived the day before the wedding, and Haefner worked all day to get the bouquets, buttonholes and venue decorations ready for the following morning. 'On occasions when I just can't get the work done the day before the wedding, I look at what I have left to do, estimate how much time it will take and then get up extra early in the morning to start working,' she says.

When President John F. Kennedy announced in 1961 that the USA would put a man on the moon by the end of the decade, many people thought he had over-promised. Instead, he sparked motivation and commitment, which led to the successful Apollo 11 mission. Promising an end date at the outset paid off for JFK, but generally it is a risky strategy for managers: however fantastic your project, it is unlikely it will create as much loyalty and enthusiasm as a moon landing or have access to the same kind or volume of resources.

Watch out for these types of project. Whatever the reason for the time constraint, the planning approach for time-bound projects should start off in exactly the same way as for those without time constraints. Work out the end date, ignoring the fixed delivery date, even if you do have to crash your analysis time into a shorter period and use estimates with a greater degree of uncertainty than normal. If your schedule shows that you can deliver before the expected date, that's great. If not, move into proactive planning mode and get creative to find a way to deliver on time:

- Run tasks in parallel instead of sequentially to save time. Look at Chapter 39 to find out how you could make your critical path work more efficiently.
- Who on the team is overstretched or overallocated and would benefit from more resources? Ask for additional people to assist with the tasks where they would add the most benefit and help activities finish earlier.
- Work overtime, and get the team to do the same. Negotiate pay rates for overtime: don't be penalized for someone else's poor organization.
- Postpone training courses or holidays, although be prepared to pay for this in the later stages of the project or during their next piece of work.
- Make decisions faster. Wheel out your sponsor earlier to speed up the decision-making process. Stop going to committees for decisions and become a 'just-do-it' team.
- Plan backwards from the end date. It gives you a different view of the project tasks and might make opportunities for time saving more obvious.
- Cut out bureaucracy. If it normally takes a week to get a document signed off through the normal process, walk round to each signatory's desk with the document and collect the signatures in a morning.
- Look again at quality. Would the project be delivered faster if the quality criteria were lowered? For example, it might take a lot longer to design and build a website that has response times of less than a fraction of a second. If half-a-second response times were acceptable, the website build would perhaps take less time. If necessary, another project or a newly formed phase 2 could improve on the response times later.
- Draw on your admin support team. If you're lucky enough to have a project support office, offload some of the administrative or routine project management tasks to them. You don't? Then procure your department secretary or your boss's PA for jobs such as minute taking in your meetings and typing reports or even an extra body for testing.
- Call in some favours to get things moving more quickly.
- And if all else fails, use your influencing skills to negotiate an extension for the delivery date.

WARNING

Throwing additional resources at a project can speed it up, but beware the law of diminishing returns. 'The traditional method for varying the duration (and the cost) of an activity is through the allocation or removal of one or more resources for that activity,' write Richard Deckro and John Hebert in

Computers and Industrial Engineering: '[This] is precisely the condition under which the principles governing diminishing returns may occur.'[69] There comes a point where allocating additional people to work on a task actually makes the output slower, and you will have to identify this point. Extra resources can have a knock-on impact on the overall productivity as well. New people take time to get up to speed, and someone has to help them during this period. It distracts the team from their own work, slowing down overall progress on the project.

GOLDEN RULES

It is rarely impossible to deliver on time, given the right amount of resources, an unlimited budget and a tightly controlled project scope, but projects seldom meet these criteria. With fixed-date projects:

- plan creatively to slash time out of the schedule;
- get the support of your sponsor for when you have to steamroller through the office bureaucracy.

38 Have short tasks

Small projects have the added bonus of delivering benefits more quickly, thus ensuring political and financial commitment from your stakeholders for the next phase. If they can see immediate results, then they are less likely to start thinking about the new solve-all project or divert your funds elsewhere. The same goes for short tasks within a project.

Short tasks provide a greater degree of visibility for progress, both helping the project's momentum and demonstrating to your stakeholders that the project is actually achieving something. The shorter the task, the easier it is to estimate accurately, meaning you can have more confidence in your overall plans.

a

BREAKING IT DOWN

When Sharon Campbell organizes a conference, she breaks down what needs to be done with the help of a project notebook. 'I have a page for speakers, a page for exhibitors, a page for publicity and so on,' explains Campbell, a conference organizer for the South Eastern Colorado chapter of the American Hearing Loss Association. Conference planning involves multiple strands of activity. 'The venue is always the most critical,' she says. The association plans events for hard-of-hearing people and those who work with and care for them. As hearing loss is a factor in other medical conditions, such as brittle-bone disease, it is essential that the venue is equipped with wheelchair access, induction loops and other services. Campbell has links with several venues and tries to book early to make sure she has the choice of dates. 'I have to look at what else is happening,' she continues. 'We avoid Memorial Day weekend and Mothers' Day, as turnout will be low, but we try to arrange things so that it falls within Better Speech and Hearing Month.' In 2006, the conference was organized the day after an assistive technology event finished, so that educators and postgraduates could stay on for an extra night and gain continuing education credits by attending.

Planning the annual conference starts nearly a year in advance, as Campbell, from Colorado, approaches exhibitors and sponsors to secure funding. 'The associations and companies we target plan their budgets at the end of the year for the forthcoming year,' she says. She starts contacting relevant groups in the autumn for the best response and documents everything in her notebook. 'You can carry paper around,' she adds, but also confesses to emailing herself copies of important documents to make sure everything is backed up.

The publicity campaign breaks down into further pages of Campbell's notebook: print, TV, radio and leveraged publicity. 'Leveraged publicity is where we find groups with similar interests to ours and target them,' she says. 'For example, there are a large percentage of retired military personnel with hearing loss, so I approach associations and magazines that specifically target that group.' By breaking down the organization into smaller chunks, Campbell can keep on top of what was last said to which speaker and which magazine has agreed to carry which advertisement.

In your plan, aim to have activities with a duration of no longer than a week. This makes it easier for:

- you to manage the plan;
- you to spot slippage and monitor progress;
- your team to manage their activity;
- your team to estimate the length of their tasks.

Breaking down a large activity into smaller tasks encourages those responsible for the work to really reflect on what is involved. This should lead to more accurate estimates and will certainly make it easier for you to identify when an overall activity is going off track. You avoid the risk of a three-month task failing to deliver on time and needing another five weeks to catch up, delaying the start of other tasks and eating into any contingency you may have had in your plan. It is a lot easier to replan and redress any slippage when a particular five-day activity fails to complete on time.

HINT

In large teams, teams split over manylocations or simply teams where you do not have the opportunity to discuss the project on a regular basis, short tasks can help promote communication within the team. To monitor progress against the schedule, you will, as a minimum, have to contact each person currently involved in a task to find out how it is going. With short tasks, you get more opportunities to interact with the team and more opportunities to build a relationship with them. It is much harder to do this if you have a conversation about project progress only every six weeks.

Breaking down the activity into short tasks will give you potentially hundreds (if not thousands) of discrete pieces of activity. If you were building an apartment block, you would end up with tasks such as 'Paint the walls of apartment one', 'Paint the walls of apartment two' and so on. Another option to consider is project segmentation, where repetitive similar activities are grouped together into segments that can be managed more or less independently. 'Network-based project management techniques [such as critical path analysis – see Chapter 39] are difficult to use when most project activities are very long compared to the length of the project, and precedence relations among activities are not simple,' says Avraham Shrub, a professor at Tel Aviv

University's Department of Industrial Engineering.[70] Building an apartment block is the type of project that might find network-based planning difficult because each apartment is worked on individually. The paint-the-walls task will have a duration almost as long as the project itself.

DEFINITION

Project segmentation, according to Shrub, is 'the division of the project work content into segments according to managerial considerations, as opposed to a division based on the type of work to be performed'. Each department requires a similar group of activities performed in a similar order, but each activity will be performed frequently and may run consecutively for each apartment or in parallel. Once the segments are defined, the duration of each task and the relationship between the tasks can be defined too. You can manage the building of each apartment as a mini-project, which gives you more flexibility in handling the plan and makes it easier to see progress at any given time.

Even projects that do not have repetitive elements that could be handled as segments can be broken down into subprojects. Subprojects have all the benefits of short tasks on a larger scale, even if the scope of subprojects is mainly arbitrary and milestones purely administrative.

GOLDEN RULES

Short tasks offer the opportunity to plan more accurately, show progress, gain commitment and deliver benefits.

39 Understand the critical path

The critical path is the longest route through a project schedule. That is, the critical path is made up of all the tasks that must be completed on time in order for the project to deliver to the planned end date. Critical path analysis works by identifying those tasks that must be done in sequence, activities that cannot start until the preceding activity is completed. By default, it also identifies those tasks not on the critical path where there is some leeway to delay work if you need to reallocate a resource to help catch up on lost time on a critical activity.

DESIGN TO DEADLINES

A project to launch a new product at Caroline Songhurst's food-manufacturing company involves people from all over the company. 'My role is to coordinate the product design and to approve the final packaging,' she says. 'When we start a product launch, the project manager explains the critical path, and we all know where we fit in.'

The critical path runs to a very tight deadline. 'Supermarkets are expecting deliveries on the day we tell them; it hurts our credibility if we don't deliver the goods, and leaves gaps on their shelves,' Songhurst, an assistant brand manager, explains. 'I work with design agencies and I can pass our internal deadlines on to them. If I don't hit my dates, the packaging won't be ready on time and the project misses a critical milestone. I don't want to put my colleagues in a situation where they have to make up time because I didn't do something quickly enough, and I don't think the other team members want that for each other either.'

For each product launch, the project manager explains to the departments involved how their role in the project affects the other teams. 'I can see how I fit into the whole project, and what my part means to the others,' Songhurst says.

Knowing the critical path helps Songhurst do her job because it gives her a view of the overall project and realistic target dates. 'The critical path for me is the most important part of what the project manager shares with us, apart from, obviously, the objectives of the product launch and what we actually need to do,' she says. 'The objectives for each launch are pretty similar, and I know my role in getting a new food to market. Those points are important, but for me the dates are the thing that decides how I organize my work.'

Critical path analysis results in a diagram representing the project. It starts in a similar way to producing a project schedule: with a list of tasks, a clear idea of how long they will take to complete and an appreciation of the order in which they need to be done. Table 39.1 shows a simplified list of the tasks required to start up a collection scheme for recycling glass.

TABLE 39.1 *Task list for project to start up a collection scheme for recycling glass*

Task	Description	Dependent on	Duration (days)
1	Sign contract with recycling plant	–	2
2	Buy lorry	1	5
3	Recruit driver	1	7
4	Distribute leaflets to residents	1	2
5	Distribute collection boxes to residents	1, 2, 4	3
6	Make first collection	3, 5	1

Figure 39.1 shows these tasks transformed into a critical path diagram, also called a network diagram. The boxes represent each activity, and the arrows show how the tasks are linked together. Signing the contract is the first thing that needs to happen. After that, the project manager can start the process to buy a lorry and recruit a driver. Leaflets are already designed and will be sent out by post, so no need for the lorry to do that. Task 5 has to wait until the lorry is purchased, as it is needed to distribute the collection boxes, and also for residents to have received a leaflet explaining about the scheme. Once they have their boxes and the driver is hired, the glass collections can begin.

There is a number in each box referring to the number of days the task will take. Note that these are the duration, not the effort required. It will take someone two hours to stuff leaflets in envelopes and pop them in the post, but the elapsed time – real calendar time – between starting and having the leaflets arrive on doormats will be two days. It is these times that allow you to calculate the critical path, shown here with thicker lines. The critical path, being the longest route through the project with no breaks, shows the earliest possible time the project can be finished: here it is day 11. The diagram also shows non-critical tasks and where there is room for some movement. For example, if the recruiting process starts a day late, it won't matter as the collection cannot start until day 11 anyway. But if the recruitment starts two days late, this activity moves on to the critical path and the project now cannot finish until day 12.

'It is often a surprise to find out what is critical and what is not,' writes Geoff Reiss, author of *Project Management Demystified: Today's Tools and Techniques*. 'Did you know, for example, that if you decided to build a tall office building, the critical path would normally run through the design, manufacture, and installation of the lifts?'[71]

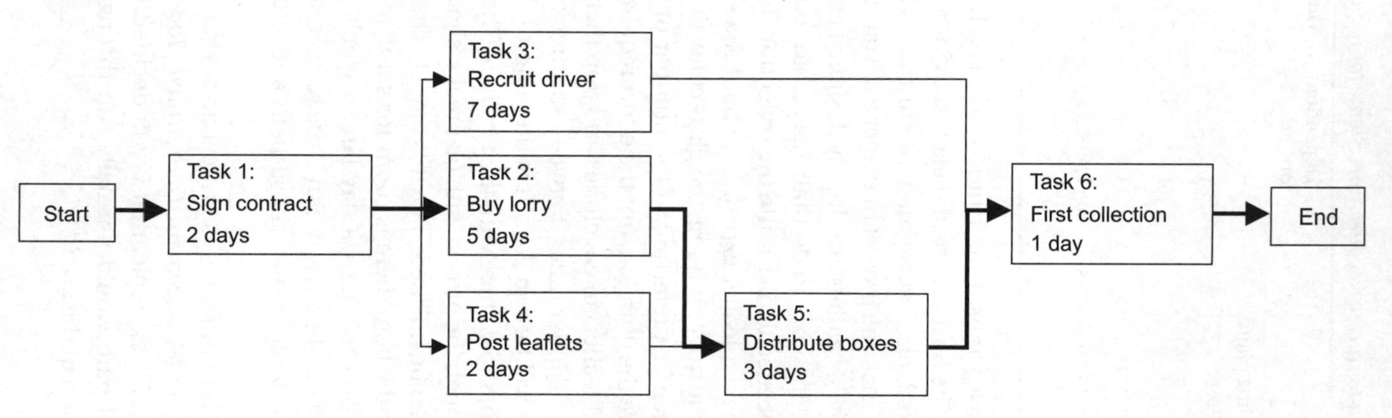

FIGURE 39.1 *Critical path diagram for glass collection project*

One of the ways to save time on a project is to look at these tasks first. Adding more resources to critical tasks and getting them done faster will definitely have an impact on your end date. Tasks will step up and become critical though, so monitor the situation to ensure you are concentrating the team's efforts in the best place at all times. Planning software will calculate the critical path for you, so with a click you can update task durations and work out what has become newly critical.

GOLDEN RULES

Critical path analysis is an aid to project planning that shows the tasks that must be completed and the order in which they must be done.

40 Understand where you're starting from

Fully understanding the current business position before you start to plan will help you identify the best order for project tasks. It will also make it easier to see where the project has critical dependencies on external activity.

ANECDOTE

The Scottish Agricultural College, an organization that supports the development of land-based industries and communities, began the deployment of SAP's financial software in 1997. Before the project started, the college was using several different financial management systems, and it was hoped that SAP would offer the opportunity to standardize processes and save costs. It was a complex software implementation, and the project team hit problems from the outset. The team did not understand fully the starting position for the project and failed to spend enough time in the early stages planning the re-engineering of the processes used by staff across the six campus locations. The team acknowledged afterwards that it would have been better to understand the dependency the new system had on the day-to-day activities of the 250 staff who would use it and to invest time early on streamlining those processes.

The project team encountered another problem later when they realized that SAP was not compatible with their own internal procurement system. They had not identified the dependency on the old system and had to invest in developing the procurement package to ensure it could interface with SAP.[72]

If you are running a project that will change the way people work – and most projects do – it is essential to have a clear idea of the starting position from which to build your plan. Earlier stages in the project may have already provided this information formally or informally, for example through conversations with your business team or through the full documentation of your scope and requirements. A solid understanding of the current position will allow you to structure your plan in the right order. Consider the following:

- Are all the business processes documented? Is there sufficient documentation around any IT systems? If not, document the status quo before trying to change it.
- Should business processes be streamlined before a software implementation?
- Is there already a team in place, or do you need to recruit?

- If you need to recruit, would it be better for the new starters to join at the beginning and help shape the project or when the project is complete?
- Will the project make some job roles redundant? If this is the case, will all the work done by the people currently in those roles be picked up by the project's implementation?

As well as getting the general structure of your plan right by taking into account the position at the start of the project, you also need to consider the project's dependencies on other factors. Dependencies form part of the environment in which your project will be operating. Understanding this environment will give your planning activity more clarity.

A project will be dependent on other activities happening within the company (outside-project, in-company) as well as on things happening outside the company and outside the realms of the project (outside-project, outside-company). Many projects will also have in-project, outside-company dependencies in the form of third-party arrangements or reliance on contractors. Finally, there are also the dependencies within the project itself, namely that one activity cannot start until another is finished (in-project, in-company). These types of dependency and some examples are summarized in Table 40.1.

TABLE 40.1 *Types of dependency*

		Company	
		In	**Out**
Project	In	Sequential or dependent project tasks, e.g. testing IT system is dependent on first writing and building it	Contractor deliverables, advertising agency copy, printing, outsourced IT deliverables
	Out	Linkages to work in other departments or on other projects, changes in company strategy or policy	Regulation, compliance requirements, health and safety standards, government or industry policy changes

Having identified your dependencies, work out their impact on your project. Some, especially the outside project dependencies, will need to be entered into the risk register. You may intend to manage others through your plan, by keeping an eye on how the situation is developing. Include any relevant people in your communications and be certain to come back to your list of dependencies regularly and add in any others as they emerge.

GOLDEN RULES

Your project will be dependent on not only your interpretation of the business situation at the starting point but also other vital factors. Identifying and understanding these will help you to plan and manage the project accurately.

41 Baseline your plan

You know that the plan you put together at the start of the project is going to bear little resemblance to the same document at the end. So what is the point of doing a plan at the early stages of a project at all? Taking a snapshot at the beginning of your project is known as a baseline and can prove to be useful later on for many reasons.

a

BASELINING COHERENCE

David Schmaltz's position as founder of strategic project management consultancy, True North, has seen him work on many pressurized projects. One assignment, a hardware and software project for a high-technology firm, found him working with, in his words, 'a dizzyingly broad community of contributors, suppliers, layers of arguing executive management, and customers'. With a seven-figure budget and a global team numbering several dozen, significant effort was put into the planning in the early days.

The team covered a wall with sticky notes to create their plan to design and deliver the blueprint for a laptop computer. 'We were baselining coherence, which allowed everyone in the community to think about the project in a similar way, whatever the plan and schedule might say.' Schmaltz is clear that the project's original plan was the result of this effort to ensure everyone involved was talking about the same thing – or, where they were not, that this potential contradiction was recognized.

A detailed project baseline was drawn up only for the foreseeable future, with the rest of the nine-month plan sketched in behind. The effect of the planning meetings early on ensured that everyone knew what was required to deliver the project successfully and also who would be responsible for each critical delivery. 'Most of the planning energy was focused upon creating relationships between the people involved,' Schmaltz says. 'We wanted to identify where critical relationships might exist and provide an opportunity for the people involved in these relationships to get to know each other, to get familiar with their working preferences, and to work together to achieve initial understanding and agreement.' He adds: 'By the end of the planning, we were reasonably confident that everyone in the room was on a similar page. And where people were on different pages, we understood that too. Everyone in the room, and, later, everyone in the community, was clear that we had not created the roadmap we would follow to a successful conclusion.'

Schmaltz firmly believes that the level of change involved in projects means they should not be a scripted performance but a fluid conversation between the stakeholders. 'The greatest danger any project faces at the beginning is mistaking unknowables for knowables, so we spent time identifying contradictions within the plan and almost no time attempting to resolve them into knowables,' he says. The team did not see any value in attempting to plan activities they did not as yet have all the details about. However, for the purpose of satisfying the funding authorities, they prepared a summary including a list of proposed milestones with a caveat stating these would probably change, given the project's environment. In the event, the project stretched to 12 months.

The baseline plan was never formally captured and used to track changes, but in practice much of it was not modified. The sticky notes were simply moved around as soon as a member of the team had better information about how a task would be achieved. 'This made the baseline a powerful medium for communication, rather than something that measured the goodness or badness of the effort,' Schmaltz explains.

The focus of the project was not on picking up and focusing on why the hardware was not delivered to the original plan, but on the end goal itself. Schmaltz was fortunate to be able to enthuse the senior management with this approach, but he acknowledges it was helped by the project's strong business case. 'I have not seen many projects that have been successfully managed by measuring deviation from initial baseline. Focusing on why it didn't turn out as planned can at best be a distraction from achieving the strategic purpose of the effort. This project, and many others like it, stayed focused on the prize.'

Setting a project baseline means taking a snapshot in time of your schedule. At the beginning of the project, all things being equal, this is how long you think the work will take. It's a target, a frozen view of a schedule, which will no doubt change as the project unfolds.

DEFINITION

Willie Herroelen and Roel Leus describe the baseline version of the schedule as 'a basis for planning external activities such as material procurement, preventative maintenance and delivery of orders to external or internal customers. Baseline schedules serve as a basis for communication and co-ordination with external entities in the company's inbound and outbound supply chain.'[73] In another study by the pair, they explain that 'booking' personnel to work on the project is another reason to have a realistic and stable baseline schedule.[74]

Your baseline schedule should be approved by your sponsor. However, this is just the first step. Just because you have a signed-off, baselined schedule does not mean that it is now set in stone. It would be unrealistic to expect it not to change at all. If you have discussed and agreed a time tolerance with your sponsor, then it will not be necessary to go back for each tiny change provided you are within your agreed tolerance. For more on tolerances, see Chapter 3.

HINT

Make sure your baseline plan is linked to a particular version of your project documents. If you're going to baseline your plan, baseline your other documents too. Your scope statement, design and requirements documents should all be under version control (see Chapter 17), so associating the relevant version to your baseline plan won't be difficult. This is useful because updates to the schedule normally result from the change-management process. A new requirement, for example, will create a need to change the plans and have an impact on your schedule. When you look back during your post-project review, you will be able to see easily how your baseline schedule was adapted following other changes.

Your baseline schedule provides a useful tracking mechanism so you can see how much has changed, which will help you monitor accurately. There are mathematical models for predicting the impact of change on your schedule,[75] but it is more likely that you will use your time tolerance, contingency, the risk-management process and common sense to accommodate changes to the schedule as your project progresses. The US Civil Engineering Research Foundation concluded in a study of Department of Energy projects that 'a rigorous risk assessment of alternative solutions under various scenarios provides a means of raising the confidence level that can be placed in early estimat estimates'.[76] They also looked at rebaselining – that is, presenting a new, adapted (and normally longer) schedule to the sponsor and having that accepted and signed off as the new target timeline – and found that frequent rebaselining 'masked the true state of some projects'. Rebaselining is possible for your projects, but always keep a copy of the original schedule, however 'wrong' it has turned out to be. There are useful lessons to be learned from why the original schedule was so inaccurate, even if for project-reporting purposes you now report on progress against the new schedule. This document will inform your post-project review and can be used to help improve estimating on projects in the future.

GOLDEN RULES

Get your schedule approved and baselined at the beginning of the project as it will help you book resources and monitor progress.

42 Record time

Project recording normally takes the form of timesheets completed by the project team on a weekly basis. Time recording is a seemingly straightforward and useful task, but it can alienate team members if the rationale behind it is not explained and the methods are applied too rigidly.

a

BLOCKING EFFICIENCY

Luke Reader has worked for project managers who handed him a list of tasks for the week. A typical list would include the hours completed for each task and the estimated time needed to finish each one already filled in. Despite the good intentions of his project managers, he did not find working with timesheets in this way very effective.

'Timesheets can put a barrier between the project manager and their understanding of what's happening,' Reader says. 'It also annoys the team by treating them as a production line rather than intelligent people who can usefully participate in managing their workload.'

Reader has witnessed at first hand how using rigid time recording can backfire and, as an IT project manager himself, has developed a more effective way of handling the work of his own team. 'The problem with timesheets on their own,' he says, 'is that the team soon learn that rather than say "This task is late, I mis-estimated", they invent new tasks such as a test cycle or a further review. These are then written on the timesheet in an attempt to show how hard they are working, even though the overall work is behind. The project manager cannot tell what the genuine issues are. And while the project manager can go back and challenge things, it means another cycle of going back to people – and time passes, which is something you don't have on projects.'

Having learnt from the mistakes of others, Reader now takes a pragmatic approach to managing his team's time. 'For me, timesheets are a mechanism to allow contractors to get paid, internal and external billing to take place, and sometimes for company management to get an idea of what their staff do all day,' he says. 'So, as a project manager, I make timesheets as simple as possible, ideally just having one task like "Work on project x", and I manage the people via discussion using the project schedule as the reference.'

Timesheets allow you to monitor the work of your team. In its own right, this is a vaguely useful activity. In a world where as a project manager you need to be able to justify what your team have been doing, and maybe cross-charge another department or organization for their time, timesheets serve as a method on which to base invoices. They are certainly not a perfect way of monitoring activity, but as no one has yet come up with an accurate and foolproof alternative, many organizations rely on them. Time recording can be used for more than just checking up on your team or satisfying internal accountants.

You can compare the timesheet data with the original estimates in the early stages of the project and see how closely they match. A discrepancy will show you that the estimates are inaccurate. 'Inaccurate estimates also cause bad decisions,' writes Curt Finch in *Projects@Work*. '"Inaccurate" usually means "too low". When this happens, the return on investment calculation shows the project as "worth it" when it is not.'[77] Doing this comparison exercise early on in the project can give you an insight into any potential overrun you may encounter in the future.

However, getting accurate data from your team can be difficult. Their timesheets may show routinely that they spent eight hours a day working on your project, but that implies they took no toilet breaks, made no phone calls and never responded to any non-project emails, which is highly unlikely.

WARNING

A project manager spends 10 per cent of their time chasing progress reports – about 16 hours a month. Team members spend 12 hours a month reporting their progress. For a medium-sized company, this equates to about £600,000 wasted a year on tracking down information.[78]

It is easier, then, to do your comparison at a stage or phase level rather than a task level. It is also essential to explain how you're using the data to encourage your team to be realistic. Frequent discussions about the project schedule will help your team feel confident that you understand what they're doing. If they feel strongly that you appreciate and are responsive to their difficulties at a task level, then this will create less of a reason for them to submit inaccurate timesheets, especially as they know you are aware of any discrepancies.

GOLDEN RULES

Time recording can be a useful activity, but don't let it be the only way you monitor your team's activity, otherwise you risk being fed inaccurate data week after week.

43 Make meetings productive

Nobody wants to sit through a meeting and feel it was pointless, so why is up to half of the time attendees spend in meetings wasted?[79] It demoralizes your team and is unproductive. Putting some thought into the objectives and format of your meeting in advance will allow you to get the most out of your attendees' time.

a

SAVED BY THE BELL

Starsys Research, an engineering company that provides mechanical systems to the aerospace industry, had only eight people in 1991. But the team was not communicating and was missing project deadlines because individuals were not aware of each others' needs. The company instigated a daily meeting to bring the employees together and manage the tasks for the week. They set up a whiteboard and used it to record actions.[80]

'As a small aerospace company, Starsys Research Corporation provides a microscope into processes such as project management. Things happen on a scale where the process is visible, and it's fairly clear where things are working and where they go wrong. We have a great opportunity to watch the programme-management process closely and find out what does and doesn't work,' says Scott Tibbitts, president of the Boulder, Colorado, firm.

Peer pressure encouraged the team to put their own actions on the board, as everyone was keen to demonstrate that they were hard at work. Although it was useful, the meetings could drag on. The senior team decided to limit the meeting to 15 minutes by ringing a bell at the start and end. The pressure was on to finish on time, and they realized they could update progress on more than 50 actions in less than 10 minutes.

Starsys grew to 140 people, and the morning meeting evolved along with the company. 'We no longer track the actions: that became impractical at about 75 people,' Tibbitts says. Instead, group leaders took it in turns to present on a subject of their choice at a series of seminars that ran until the spring of 2005. Peer pressure was still used to keep the meeting productive: after the presentation, a vote decided how useful the session was. The department whose presentation was ranked the most useful each quarter got to go out for a meal.

Before you cancel a regular meeting, ask yourself why the time has become non-productive. If a meeting would truly be irrelevant to all the participants, then cancel it: do not waste people's time with a pointless get-together. But if you believe that arranging a time for everyone to be together would genuinely be beneficial if only you could increase the collective productivity, then think what you can do to make the meeting more efficient. Your office may not have the space for a whiteboard to list all your actions, but if you would like to be able to squeeze something useful from the time, try some of these suggestions:

- Make sure people know why they are there, so they can plan their contribution in advance: send out an agenda and objectives for the meeting. In a research study done on engineering projects, Ana Garcia and colleagues found that the agenda was the key to an effective meeting.[81] Allowing anyone to suggest agenda topics can mean that the final version of your agenda contains lots of items that are of no relevance to the majority of attendees. The researchers conclude: 'An ideal agenda contains only items that need the attention of mostly the entire group . . . purely informative items would be better dealt with through other means of communication. In addition, issues that concern only a few people in the group should also be discussed in another forum.'
- Take it in turns to be the chair: this method also allows quieter members of the team an opportunity to speak and take the lead. However, keep the role of secretary yourself, as you need to be confident in the completeness and accuracy of the minutes.
- Do not assume action points from the previous meeting have been completed. Go through the minutes, ask for updates and, if an action is not done, carry it forward.
- Set a firm start and end time, and consider imposing penalties for those who are late. Starsys asked for financial contributions towards a team party from latecomers.

HINT

The guidelines above still apply for a meeting held 'virtually' by video,web or telephone conferencing. However, because the interpersonal relationships and reactions between your colleagues will be harder to understand and respond to, you need to apply an additional set of rules to virtual meetings.

Steve Draper, a researcher in the psychology department at the University of Glasgow, was part of the project team that developed a new method of collaborative teaching and tutorial support using two metropolitan area networks to link four Scottish universities. Part of the teaching was done using video conferencing, and Draper and his colleague Margaret Brown documented some important tips as a result of the implementation:[82]

- If your video link is not sophisticated enough to present several screens, set up a parallel computer link to display presentation slides or make sure attendees have a copy of the presentation on paper in front of them. 'Audiences say they quickly get tired of hearing without seeing the speaker,' Draper and Brown conclude. Don't rely on keeping your audience's attention if you point the camera at the screen of your laptop.

- Send out an agreed agenda in advance, and use this email to introduce yourself if you are unknown to some of the participants.
- Practise using the controls for that particular video-conferencing suite in advance. It is likely that they will be different to any you have used before. Fumbling makes you look unprofessional, and the attendees will be left with the impression that the meeting was a waste of time. The same goes for web-conferencing software: have a dummy run first so you know what you're doing.
- If you can get someone to turn up early at the other end, practise the camera angles and make sure they can hear you.
- During the conference, keep checking that the communication is working: 'You must ask the other end and believe what they say,' Draper and Brown advise in their document. 'The fact that you can hear OK is, unlike in face to face, no clue at all about what they can hear.'
- Subtlety is not going to work. 'If their sound is too quiet, it is no good talking louder. They hear fine and won't talk louder to suit you, particularly if they have several people in their room who can hear each other fine,' the researchers write. You will have to explicitly ask your team members to speak up and emphasize your body language to compensate for poor resolution at the other end.
- Don't just do it for the sake of it. Telephone, web and video conferencing can add variety to your project team meetings, and in that respect are probably worth trying, but weigh up the advantages and novelty factor against the potential downsides. Draper and Brown conclude: 'After all, video conferencing is inferior in many ways to face-to-face meetings (e.g. no social or private business with others "on the side"), and must have a strong saving in time to be worthwhile.'

GOLDEN RULES

You can fit a lot into an hour-long project team meeting if you plan in advance and brief your team as to what to expect. Plan for productivity and you will probably be surprised at the results.

44 Delegate subplans to workstream leaders

On large projects, your plan will include hundreds, if not thousands, of tasks. Having all those in one schedule document is going to make tracking really tricky. You can get round this by delegating the management of subplans to your workstream leaders and having only high-level milestones in your overall project schedule.

ANECDOTE

When Claire Simpson was asked to run a project to refit a 60-foot corporate yacht, she knew she was looking at a schedule that would run to pages and pages of tasks. 'Our company specializes in refitting yachts, and from the work that needed to be done on this one I could tell it was going to be a tricky project anyway,' she says. 'I split the work up into chunks: there were tasks for the electricians, our IT technicians were involved for the onboard equipment, engineers needed to overhaul the mechanics and we were completely replacing the deck, so that involved mainly carpenters, managed by our design architect. In all, I think there were about six different strands of work.'

Simpson nominated a workstream leader to manage each strand – the senior team leader in each department – and asked them to prepare a plan. Once these were ratified and the dependencies between each plan agreed, she produced an overall project plan and associated schedule noting everyone's key milestones.

'Managing the plan like this was easier,' Simpson says. 'We got together weekly to discuss progress, check where we were in terms of the client's schedule and just reassure ourselves that it was all going to plan. There are a lot of dependencies in yacht refits because the area we are working on is quite small and we can't have every team piling in at the same time.' Simpson says she always works with workstream leaders, and it is a technique that helps manage contractors too. 'We don't have a permanent sail maker on staff, so if a yacht needs repairs or new sails, we contract the job out. One of the workstream leaders can manage that relationship,' she explains. What is important for her is to have absolute trust in the workstream leaders. 'We've been working together for years so I know that if they say a job is going to be done, it will be done,' she says. 'We can't do this job if we're not a team.'

DEFINITION

A workstream leader is someone who manages a discrete group of tasks or people and reports to you. For example, you may have a project that involves input from the IT, marketing and customer service departments. Each department nominates a coordinator – a workstream leader – to organize the work their area needs to do. They can be a great help: they will know the people involved better than you and can select the right staff, provide guidance on internal politics and how long work going through their department will take, and generally be your expert in the field. An example of a project organization structure is shown in Figure 44.1.

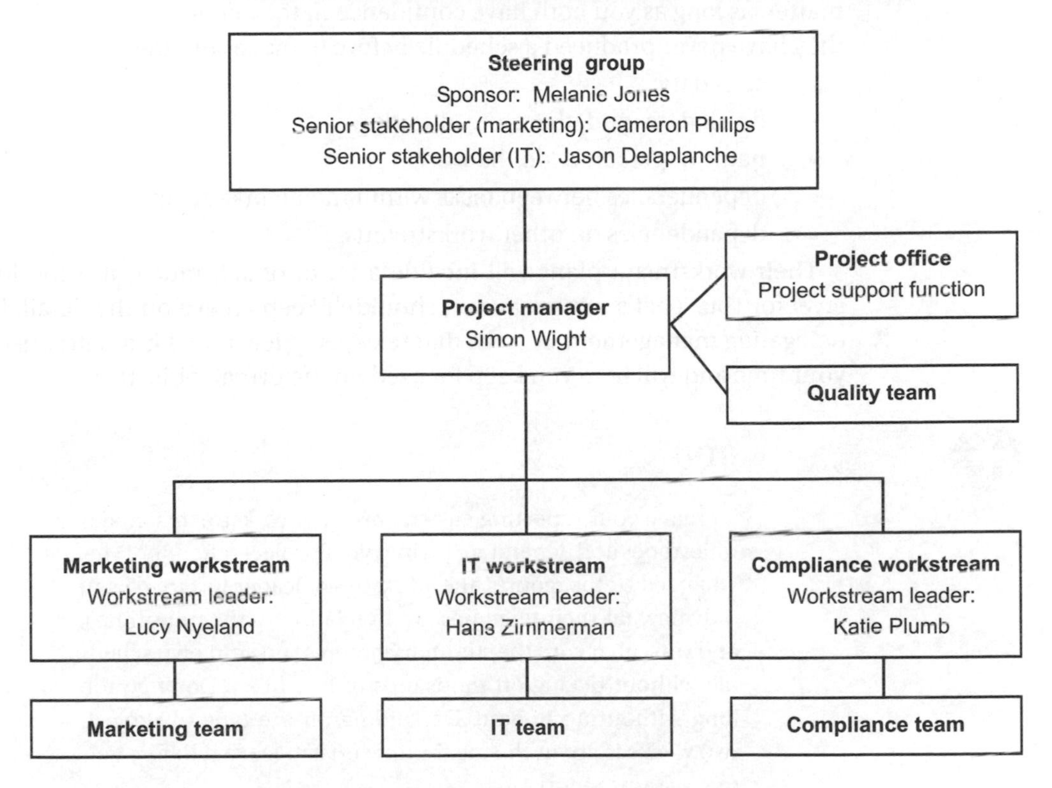

FIGURE 44.1 *Example project organization structure*

Once you are aware of the different tasks required to deliver your project, the workstream leaders can start to plan their areas in detail. Planning, in its widest sense, includes documenting all the deliverables and working out the order in which things go. For their workstream plans, your workstream leaders probably won't feel the need to write down and describe anything in a very structured way, as they will have a clear idea by now of what you need them to do. You can encourage 'proper' documentation if your internal planning methodology requires it and you believe it would be useful for everyone, but the things that are really important for you are:

- the schedule – the list of dates saying what gets done when;
- resource usage to feed into your overall budget estimates;

- constraints that will affect the way they work;
- any assumptions they have made;
- the prerequisites they require before they can start;
- any dependencies their workstream has on other project activities.

You can then add any workstream-level risks to your risk register. These details will also help you have a cohesive view of all the activities and how each workstream interacts with the others.

Whether your workstream leaders produce their schedule using a planning tool such as Microsoft Project or on the back of an envelope doesn't really matter, as long as you both have confidence in the end result. Guide them if they have never produced a schedule before to make sure they include:

- start date;
- finish date (and, therefore, duration);
- name of person to do the work;
- dependencies between tasks within their workstream;
- dependencies on other workstreams.

Their workstream plans will include a lot of detail, much of it too low-level for you. That's not to say you shouldn't keep an eye on the detail, but delegating management of tasks that take just a few hours is a better use of your time and will help you keep focused on the overall objectives.

HINT

To make your reporting easier, copy the workstreams' major milestones and dependencies into your project schedule. When you need status reports, the workstream leader has to tell you only how far their team has got in relation to the milestones, and you can record the summary progress on your own schedule without ticking off thousands of tiny tasks. Don't go too long without an update. Depending on the type of project, two weeks is about the maximum you can leave it before getting a status report.

GOLDEN RULES

Delegate the management of subplans to workstream leaders to keep your overall project schedule a summary of all the activity across multiple teams.

45 Work out when you will leave

Project management is, by nature, a transitory job. You work on a project, it is completed and you move on to something else. But working out when to leave is not always as clear as that. Some projects just seem to go on and on, with no end in sight.

a

MOVING ON

Jackie Garvey, a New Zealander who has managed projects in France and the UK, says: 'When the different parties start arguing among themselves, you know they have taken ownership and it's time to leave.' Garvey was on the steering committee for the launch of a web-based portal for her company's staff. She provided programme management and coordination from the company's global headquarters while a local project manager led the implementation itself. As the two-year project came to a close, it was clear that the local IT and business teams had accepted the portal as their own: the heated debate about its future made it obvious to Garvey that her guiding role was no longer needed.

She gives another example: 'I was working on the creation of a new IT company to manage our hardware and IT infrastructure,' she says. 'Once that project was complete in one country, we wanted to use it as a global standard. That meant the launch of a new company headquarters and setting up the new organization in all the countries where we operate. At the end, I knew the operation inside out. I was too close to it and it was time to go.'

Projects have, after all, a start, a middle and an end. Being able to plan your exit is an important part of the end. As you will have spent a fair amount of time working on the project, you will be widely considered as a (if not **the**) recognized expert, and even after the work is completed people will still direct their questions towards you. Unless you want the project to turn into your new day job, you will need to invest some time in planning how you will bow out gracefully. In the last phase of the project, it may become obvious that the time is right for you to move on.

The first thing to do if you find yourself wanting a change after working on something for a long time is to talk to your manager about moving on to a different project, perhaps with a consultancy role in your current project until the new project manager is up to speed. In some cases, there will be no choice but to stay to the end, so a bit of thought as the project comes to a close will help you develop a watertight exit strategy. Think carefully about what you will need in place for the project sponsor to be happy to let you go. This may include getting the business-as-usual team to shadow you, organizing training and preparing handover documentation.

A natural end is after the post-project review. A project manager can chair the post-project review and circulate the minutes as their last task. The 'project' as an entity in its own right may end there, but the implementation is (we hope) there to stay. There will be ongoing tasks to do. One of these is benefit tracking. If you will not be around to monitor the project benefits, make sure that the person who will be tracking knows what is expected from them, including how to use any model you have set up.

Once you do decide to move on, do it quickly and become involved in something else as soon as possible. An in-between period will inevitably mean you get sucked in to the daily routine of your ex-project. Make a clean break and don't be afraid to bounce questions back to the business-as-usual team. They are the recognized experts now, not you.

GOLDEN RULES

Make your exit an inherent part of your project planning. When you have decided it is time to move on, go quickly and with a full handover to the operational team.

Section 5
Managing yourself

Personal influence is not to be trusted beyond a certain limit.

Caius Cornelius Tacitus (c.56–c.117), *The Histories, Vol. II*

Researchers can't agree about what makes a good project manager. Each new survey produces a different list of the top 10 characteristics for successful project managers, which just goes to show that a 'model' project manager does not exist. Every individual approaches project management in a different way, bringing with them a unique set of skills and experiences that they devote to getting the job done in the best way they know how.

This section cannot teach you how to be 'better', but it can give you some ideas for developing yourself, especially in time management and career progression. It also presents some examples of when it all goes wrong – project managers in difficult situations and how they rose to the challenge.

46 Get organized

It is a project manager's job to organize everyone else, and you will be much more efficient at doing that if you can keep on top of your own activities. If you are clear about what you have to do next, it will make it easier for you to organize other people and the work of your team.

a

ON TOP OF THE LIST

Former corporate finance executive, Tina, from Montana, found the best way to manage her daily activity was to keep a list of all the tasks she needed to complete. 'I kept my list on a plain legal pad. A lot of my co-workers were really getting into PDAs and computer organization software at the time, but they were always tinkering and it took them 10 minutes to get the thing to record the task. My paper method would require me to rewrite the list on a clean sheet every once in a while as I got through stuff and added things.' Tina found a use for all the old lists as well. 'I kept all my old to-do lists in a file in my desk for use when I prepared for my annual review,' she says. 'After a year, who can remember all the little things they've done that ended up making a big difference? Keeping those old to-do lists was a huge help, especially when asking for a raise.'

Tina has some advice for people who want to use this method: 'The importance of the list is keeping one list – not a bunch of sticky notes on your monitor, not a thousand little scraps of paper. If I had to remember to make a phone call, it was on the list. Meet a deadline, mail something, call the doctor before 5pm, get back to someone, whatever . . .' she says. 'Giant projects sat on the list, like "corporate budget", until the project was undertaken, and then I would break it out into the thousand pieces that had to be accomplished.'

She found a great deal of satisfaction in drawing a line through a completed task, and realized that her to-do list also provided an easy way to demonstrate to her managers what it was she spent her days doing. '"The List" is a great way to show or, for some people, prove how busy you really are,' she says. 'There is a lot of political capital gained with your boss when he or she sees you pull out The List to discuss what you're working on, especially if The List is long.' Tina is sure that having a list of activities helped focus her attention and make her more productive. 'The List Game became an endless cycle of me trying to find a way to get everything off the list,' she confesses.

Keeping yourself organized is a step towards being more effective and being able to organize the others in your team. If a paper list to organize your activities is not going to work for you, then experiment with some other methods until you find one that does:

- Have a separate section on your project plan for project-management tasks, and use that to keep on top of your activity.
- Use an electronic handheld device.
- Tuck a list in the back of your Filofax or write the tasks in the calendar section of each day, carrying them forward if you do not complete them.
- If you can hold the list in your head, practise mentally juggling the order until the most important items are at the top without forgetting the smaller jobs.

Knowing **what** to do is only half the challenge. Knowing **when** to do it is almost a separate skill. Being able to prioritize your activity is another trick to get the most out of your time. Take your list of tasks and work out for each one its importance and urgency in relation to the others. Some tasks can be incredibly important, but not very urgent – at least for today. Figure 46.1 shows the four categories of prioritization. Both identifying tasks and prioritizing them are things to be done on a very regular basis, as projects shift on a regular basis and what is needed urgently for tomorrow may suddenly become less urgent, giving you more time to focus on something else.

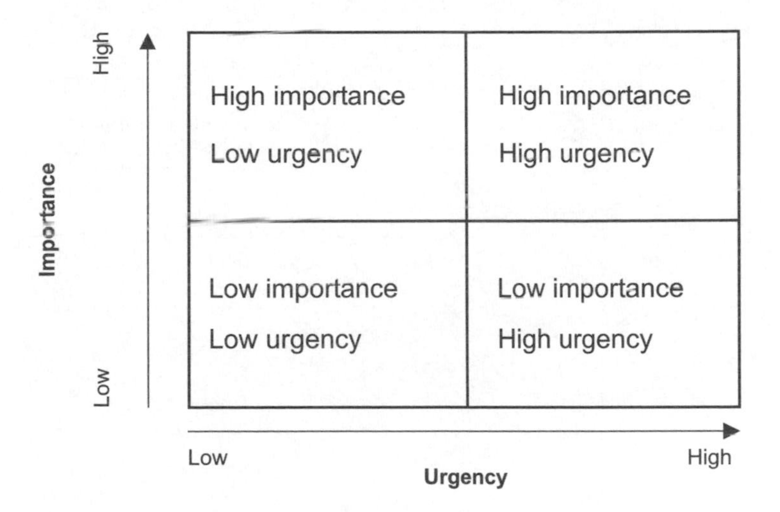

FIGURE 46.1 *Urgent and important tasks*

HINT

If you are struggling to keep on top of your list of things to do, and find yourself with not enough time in the day, make your phone calls first. A call before coffee break in the morning will not take as long as the same conversation at the end of the day. In the morning, everyone is busy sorting out their day and will want to get the conversation over. At 4.30pm, people are starting to look forward to going home and will drag out the conversation.

GOLDEN RULES

One of the crucial skills of a project manager is organizational ability. Find a way that works for you to keep on top of the tasks that need to be done.

47 Keep your records tidy

Project management produces a lot of paperwork. As well as the official documentation, there are also meeting notes, reminders to yourself, scraps of paper where you have scribbled budget changes or useful telephone numbers . . . the list goes on. Keeping those piles of paper in a semblance of order is important, as it means you can instantly find everything you need.

a

THE PAPER PILE-UP

Diane, a charity worker, held several key positions on different boards, all of which dealt with women-related issues. She was a well-respected, hard-working person. But she struggled with information overload, which was affecting her ability to do an excellent job. She sought advice from Barb Friedman, a professional organizer and director of Wisconsin-based company OrganizeIT.

'Papers were everywhere: in the kitchen, in several briefcases, next to the computer, on a desk and in her bedroom night-stand,' Freidman says. 'She couldn't find information when she needed it and spent countless hours retracing her steps looking for it. Her stress level was off the charts.'

Diane chaired two of her boards, and it was essential that she could lay her hands on information in a hurry. Friedman explained three principles of staying organized to teach Diane the skills and process so she could learn to maintain the system by herself. 'We sat down and discussed prioritization, categorization and storage. Once we began streamlining papers, the topics start to create themselves. I set her up with paper systems that are easy to use and easy to maintain. We used lots of binders—one for each of her board positions, one for the history of the project and so on.'

The approach was certainly successful. Friedman called Diane a few months later to check she was still managing her filing. 'She was so relaxed and upbeat,' Friedman says. 'She said that while at a board meeting, a discussion took place regarding a past decision. She was the only board member to find the necessary information in her binder.'

Project managers are largely left to their own devices when it comes to deciding on a method of organizing their paperwork. Once settled on a routine – good or bad – it is easy to become stuck, finding workarounds to a poor or complex filing system because there is no time or inclination to properly review and revamp.

UP IN SMOKE

Project manager Celine Montague was forced to review her approach to storing information when her PC suffered a terminal crash and she lost all her project documents, which literally went up in smoke. 'All my notes for our European fraud project were on my hard drive and not backed up,' she says. A freak power surge at Montague's Brighton-based company put a handful of computers out of order. She realized what a mistake it had been to not have a backup when a curl of grey smoke drifted out of her PC and had her racking her memory to recall critical project dates and decisions. 'Since then, I've always burned backups to CD and kept paper copies of really important documents,' Montague says. 'I learnt the hard way!'

Your filing system needs to be something that works for you but that is also time efficient. Table 47.1 offers some tips of how to get started. Turn to Chapter 56 for more on long-term archiving after the project.

TABLE 47.1 *Tips for record-keeping*

Do	Don't
Categorize your filing: have different sections in electronic or paper filing systems for different things	Write out notes by hand and then type them if you can help it: it's a duplication of effort. Instead, take notes directly on to a laptop if you can
Prioritize: have frequently-referred-to files at the front of folders or at the top of the directory tree on your PC	Throw things away for the sake of it – sort them out first
Date all documents, even scraps of paper	Forget to back up your electronic records
Stick to it!	Make filing so complex you forget how you set up the system
Take good ideas from your colleagues	Feel you can never change how it works for other projects
Put papers in an in-tray if they are waiting to be filed, and file them before the pile gets too big	Use your email inbox as a filing cabinet – save attachments on a network drive and delete the original email
Sort out your emails at the end of a project: create an archive file and store it with your electronic records	

GOLDEN RULES

Keeping your project records tidy means it is easier for you to find information and also to be clear about what you are missing.

48 Don't lose sight of the end goal

It is easy for a project manager to become so engrossed in the details that they fail to see the bigger picture and spot when a project is losing focus. It is even easier for the sponsor or other stakeholders to lose sight of what they originally wanted to achieve with the project and change it beyond recognition and beyond the project manager's control. Managing your scope normally means understanding what your customers need and, once that is agreed, keeping a tight rein on change control. The most efficient change-control process known to project-manager-kind will not help when you or your steering committee lose the vision of the project as a single project. Try to be aware of the end goal and keep the project objectives in mind so you can identify whether poor focus is likely to affect the judgement or actions of you or your sponsor.

a

WHEN SCOPE CHANGES GOES WRONG

Dutch healthcare rules are complex, and companies have to adhere to strict legal procedures when employees are away from work during an illness. Employees are able to receive government assistance for long-term sickness after two years out of work, but at that point, if their employer has not undertaken all actions to get the employee back to work and kept excellent records of its actions, the company risks a serious fine.

When the laws were changed recently, one Dutch company realized it opened up the opportunity to offer a service helping companies with the administrative duties while supporting employees' treatment and getting them back to work as quickly as possible. The product was an innovative idea and one the company wanted to implement quickly. 'The idea behind the project was brilliant,' explains Paul van Doorne, who led the project team to implement the IT system that supports this new product. 'We would leverage our existing expertise and partner with a private disability product company to enter the market. We looked for a relationship with an experienced service provider who would be able to take care of the interventions to get the employee back to work and the administrative support. The combination on paper looked like a winner.'

As with all new products, there was no guarantee that it would be a success, however promising the initial business case. Van Doorne, an experienced programme manager leading a team of process improvement specialists in the Netherlands, negotiated the setup of a separate back-office system to support the early launch of the product. 'The existing back office had problems with their IT systems,' he says. 'They weren't able to support the processes needed for the new product launch. We decided to keep the scope tight – just to build a solution to support the launch.' Once the product had proven itself, the operations department could be reviewed as a whole and all their IT systems reassessed. There was another driver for an initially simple solution: the steering group agreed a business case that put operating costs at 30 per cent lower than the costs of the existing organization, so there was pressure to keep down the development costs.

'We selected a simple software solution to support the product and started the implementation,' van Doorne says. 'The project steering committee was aware of the rationale behind our choices, and approved them: we would only use the simple solution for six to nine months after the launch, which would buy us time to think about the long-term solution – probably a new IT system and re-engineering the operations department. After nine months, we all expected to throw the software away.'

A few weeks into the project, the steering group started using their regular project meetings to discuss the general problems in the operations department. 'Those issues definitely should not have been a topic for the steering committee in their role as part of a defined project,' van Doorne says. 'I think they suddenly saw the potential of the process being run at 30 per cent of the costs and suggested that we move all the products into the new system. Keep in mind that our aim was to use the software for a maximum of nine months, so we just hadn't built in the functionality for things like renewing annual contracts,' he adds. Nevertheless, the steering committee asked van Doorne and his project team to first convert all the company's existing products to the new system and then work out how to add in the functionality in order to ensure the system would work efficiently in the long term.

'In my own experience, and that of the managers who work for me, I've known many projects where a sponsor or steering committee has seen the project manager as an assistant for the normal business. But on this project it meant a real expansion of the scope,' van Doorne explains. The project leapt from a new product launch to a full review of all back-office functionality and implementing a complete new IT system.

'The normal reaction of project managers on scope changes is to calculate the consequences in time and money and then ask for approval of the expansion,' van Doorne says. 'This time, I did the same but also emphasized the fact that we had selected a software solution that would only support the launch of our amazing new product.' Unfortunately, the steering committee had lost sight of the original deliverables and van Doorne wasn't able to bring them back to it. The steering committee consisted of a number of directors but only one member of the operations department. 'This scope change also created the strange situation that the steering committee decided to make changes in the operations department and overruled the director of operations himself,' van Doorne says. 'The steering committee thought that it was a good idea for me as the project manager to take the lead in re-engineering the organization as well. It was strange to see a project get completely out of control and the senior level of the company approve it.'

After a review, it turned out that the new software solution was not able to support all the products and that adding the functionality to handle long-term requirements such as contract renewal was just not possible. 'This was exactly what the project team expected,' van Doorne says. But the experience wasn't all a disaster. Van Doorne summarizes what he learnt from the project: 'It is essential to keep the scope tight and not only defend scope changes by stating their consequences in time and money but also to reassess decisions made earlier in the project in the light of the new proposed scope.'

Although project managers have a lot of power in relation to getting a project done, there is only so much you can do to influence the sponsor: highlighting all the associated risk factors, raising issues, using the normal and exceptional communication and escalation channels. This book is scattered with examples of project managers finding themselves in situations where they have to react to poor or lack of decision making from their superiors. What is harder to do is to recognize when **you** are losing focus and becoming so lost in the minutiae of project tracking, monitoring and replanning that you fail to identify that the project is now going to fall way short of its original goals.

Keeping the end goal in mind is essential for success, however you have defined success. In *The 48 Laws of Power*, Robert Greene and Joost Elffers write:

In any organization it is inevitable for a small group to hold the strings . . . As Richelieu discovered at the beginning of his rise to the top of the French political scene during the early seventeenth century, it was not King Louis XIII who decided things, it was the king's mother. And so he attached himself to her, and catapulted through the ranks of the courtiers, all the way to the top.[83]

Richelieu identified his end goal and never once lost focus on his way to achieving it.

WARNING

There are few hints or tricks that guarantee the avoidance of lack of focus, either from the project manager or from the sponsor. Just be aware of your role and perhaps set some time aside as you do your monthly reporting to bring your attention back to the original project purpose. In your communications with your sponsor, always try to answer the question 'How does this help us achieve our objectives?'

Think of your role as a helicopter (see Figure 48.1). You need to have an overall view of all the activities that make up your project – scope, quality control, the schedule, your team and so on – to be able to zoom in and deal with any crisis in any area. But you also need to be able to pull back, hover and see how the whole picture looks together.

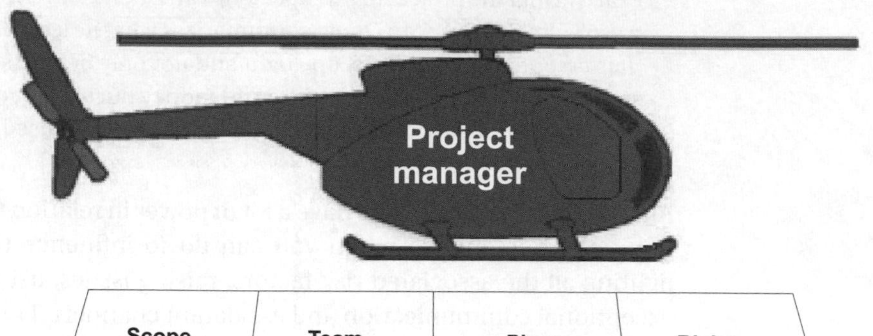

A project manager should be able to 'land' on a problem and resolve it while still seeing the big picture.

FIGURE 48.1 *Helicopter view of a project*

GOLDEN RULES

The project manager's role is to manage the project on a big picture basis. If you or your stakeholders lose sight of what you set out to achieve, you certainly won't achieve it.

49 Promote yourself

You are a highly competent, technically excellent project manager who delivers results time after time, so why are your efforts being ignored? Unfortunately, it is often not enough to do a good job. You also have to make sure you are seen by project sponsors to be doing a good job. And not only are project sponsors frequently uninterested in your activities – they also have a very different idea from you as to what constitutes a job well done.

a

PROMOTING A PROTÉGÉ

Alison Batten, a programme manager in the insurance sector, was mentoring a junior project manager who was hoping for a promotion. 'Together, we sat down and looked at the projects she was running. One was for a highly influential sponsor, whom I knew well. I knew he'd prefer to receive concise but detailed reports, and I found out that at the time he wasn't receiving any project information outside the monthly steering group reports. I suggested she sent him a summary of her more regular updates. She didn't need to change how she was managing the project – just how he thought she was.' The junior project manager changed the way she communicated with the sponsor and started to see some results in how she was perceived. 'Don't underestimate the power of having a senior sponsor on side,' Batten says. 'You're far more likely to be requested to work on their pet projects and your availability to do so is likely to be discussed at a senior level.'

At the same time, Batten suggested that her protégé created a higher profile for herself. When Batten was on holiday, she left the running of the programme in the project manager's hands, including chairing a programme meeting with several senior members of staff. The project manager found that the more people who knew her and appreciated her skills, the easier it was to put forward the case for a promotion, which she got a few months later.

'I used a similar method myself,' says Batten, 'in order to secure my last job move. Ensuring the key people know what you are capable of is essential to getting the recognition you deserve.'

Lynn Crawford, a Sydney-based researcher, studied the gap between how good project managers were at doing their job and how good their supervisors thought they were.[84] She tested over 200 project managers against both their knowledge of techniques and their practical ability to do the job, using the Australian National Competency Standards for project management. Then she asked their supervisors to rate them on four criteria: their value to their clients and to their organization, the ability to motivate others and their ability to work with colleagues to deliver a project. The results were surprising. There was no statistical correlation between doing well on the tests and receiving a good supervisor score. Crawford concluded that 'the knowledge and practices valued by project management practitioners, and embodied in their professional standards, are not the same as the knowledge and practices valued by senior managers'. When she looked deeper into the results, she found that if project managers wanted to be perceived as excellent at work by their supervisors, then they should concentrate on:

- working on projects with a high level of ambiguity in either scope or methodology;
- working on projects that differ greatly from each other;
- developing an excellent knowledge of cost, time, procurement and HR management, and using these skills to control their projects;
- not becoming too involved with general management activities, in order to avoid being seen as interfering.

In an organization where projects are allocated to you without the opportunity to choose or express a preference for your work, the first two recommendations may be difficult to achieve. Concentrate on trying to influence those people that make the decisions about the projects on which you work. If they see you starting to operate at a more senior level, then there is more chance of them giving you the challenging projects you need, which will in turn raise your profile elsewhere in the organization.

Following Crawford's last two recommendations will improve your skills and how others perceive you, but you will still need to attract the attention of senior managers in order for them to form a perception of your abilities. These senior managers may not be involved directly in your work and will, therefore, pay less attention to your successes.

Try to raise your profile subtly with a few of these ideas:

- Circulate useful articles to your project management colleagues or occasionally mention titbits of knowledge you have acquired: 'When I was at the marketing meeting this week, Karen told me she was leaving – I thought you might like to know, as it could affect the mailing you are planning.' Subscribe to magazines relevant to project management or your industry, many of which are free to qualifying applicants. Display them prominently on your desk so your team mates can see you are the sort of person who keeps your knowledge up to date. Starting to raise your profile in your team will have a knock-on effect in other departments, as your colleagues start to mention your name for the right reasons.
- Make sure people know of your achievements. Find opportunities to mention your successes, but remember no one likes a braggart. Slip your self-publicity statement in around another message or story: 'I was out at the weekend with a friend who told me her company has just lost five weeks of client records when their database crashed. It's

a good thing I made sure we've got proper backups on the project I'm working on – at least that can't happen to the sales database here.'

- Ask to tag along with someone who is attending a meeting with influential people. Either approach it as a work shadowing opportunity or create a reason why it might be relevant to your project.
- Never turn down an opportunity for internal networking – staff briefings, lunches, management question-and-answer sessions (have an intelligent question to ask).
- Be prepared for your encounters. Plan a 30-second (interesting) summary of your current activity so if you are asked what you're working on, you sound intelligent and articulate. You can use this kind of 'elevator' speech at conferences too.
- People like to think they form their own opinions, but in reality they are easily influenced by the opinions of others. If other people think highly of you, let it be known: 'The editor of *XYZ Journal* liked the article on risk management I submitted and is going to publish it next month.' It doesn't have to be the opinions of high-flyers or even named individuals either: 'I was chosen to do the analysis on the Project Lambda business case.' But say it in a positive tone: self-deprecation can make a compliment sound like you were picked because everyone else was working on more exciting and important projects.
- Become indispensable to your programme manager. Stay in the loop about what is happening at programme level and then offer to take the reins when they are away. Do a good job, but hand them back gracefully, with an excellent briefing of what happened. Programme managers will have more confidence in you if you do not become precious about your time in their chair.
- Don't be too modest. Project sponsors want to know that everything is under control, but if you sort out every issue and just reassure them that things are on track, they will just see the smooth progress and not your gargantuan efforts to stay on schedule. In your meetings, explain your difficulty but still add the assurance that you have fixed it: 'We had an issue with our new database servers; there was no space to connect them up in the warehouse. It was a bit frantic, but I finally managed to find the space by getting the team to reorganize what was already in there. They got it all connected in time for the sales team to start testing the new system.' No longer will your hard work be transparent.
- Never talk yourself down, even outside of work. Never use the phrase 'It's not that interesting.' Someone found your project interesting or they wouldn't have asked you to do it. Even if it is too complicated to explain in detail, you should always remain positive and upbeat about what you do.

You don't actually have to be the best-performing project manager in your company to do a good job of promoting yourself. Take the opportunities that are offered, create your own opportunities and build on your self-promotion skills. The trick is making sure that the people who matter think that you are doing a good job, regardless of how confident you feel in your own ability.

GOLDEN RULES

In order to become recognized for your project-management skills, make sure you, first, give your senior stakeholders what they want and, second, work on your profile within your organization.

50 Don't panic

Perhaps one of the hardest skills to develop as a project manager is that Zen-like quality that magically transforms changes, issues and disasters into opportunities. Staying calm in the face of a crisis doesn't come naturally to everyone, but panic is not a good look for a leader.

a

THERE'S A HOLE IN MY BUDGET . . .

When Julia found a £50,000 hole in her programme budget, she started to worry. She had been doing the figures for a group of projects that were going to update her company's ecommerce strategy. The worry soon became panic when she realized that it was an error in her maths that had created the deficit. She had made a mistake that had been carried through the entire budget spreadsheet over nine months, and the programme had spent a good deal more of the £1.1 million budget than the figures were showing. Julia instantly stopped approving new invoices but did not know what else to do. After losing a night's sleep worrying which projects would need to be cut back, she tackled the spreadsheet again the following morning.

Julia's eyes started to blur as the numbers began to jump around the screen. She couldn't think straight. She was concentrating on the £50,000 hole and panic set in. She couldn't correct the problem, because she couldn't work out where she had gone wrong. Without that understanding, she couldn't calculate the new balance and identify which budget lines were overspent. Was it training, consultancy fees or hardware? Julia had to tell her manager not only that the budget was £50,000 healthier on paper than in reality but also that she was the cause of the problem and she was unable to correct it or find out where the money had gone.

When Julia went to her line manager with the bad news, she suggested that an accounting expert sort out the maths in the spreadsheet and help her understand the issue so next time she would not make the same mistake. Fortunately, her manager agreed and saw it as a development opportunity for her. Both he and Julia knew there was nothing that could be done to get the money back, so they acted quickly to stop the problem getting worse.

In your project-management career, there will be times when you make mistakes. Whether you lose £50,000 in the workings of a spreadsheet or double-book a training room, the crucial aspect is how you handle it. If it has to be solved instantly, apply some creative thinking: can those delegates spend an hour doing the introduction to the course in the park across the road until the meeting room is free? Stay calm and professional and, if you think it would help, admit your error. Then get on and put it right with the minimum fuss possible. So something went wrong? If you handle it smoothly, it will soon be forgotten and the memory will instead be of your level-headedness and ability to resolve a crisis.

HINT

If you don't feel calm breaking the bad news, think of the conversation like a role play and try to fake it.

If it is not an instant crisis, you have a little longer to consider your plan of action. You might be able to get away with never admitting the mistake at all if it can be corrected quickly. If you plan to own up, make sure you have all the facts and, if possible, an action plan to get things back on track. This latter point is especially important if you cannot sort it out alone. When you present the situation to your sponsor, again stay calm and factual: 'We lost a day's work on Thursday because I failed to tell security that the team needed access to the building. I've cleared it now so we don't need to get permission for next time, and can you authorize the overtime Alex has agreed to work on Sunday to get us back on track?'

GOLDEN RULES

Even the most experienced managers make mistakes. Don't panic, put things in perspective and act professionally – even if inside you don't feel it.

51 Know what's a showstopper

A showstopper is something that has stopped or threatens to stop your project today. You will not need to be told how to recognize one: it will be clear from the panicked look on your team's faces and the unspoken question 'What do we do now?' that hangs silently in the air.

a

SHOWSTOPPER MANAGEMENT IN ACTION

Eric Spanitz has over 16 years of practical project experience, is a professor at Lake Forest Graduate School of Management in Chicago on their MBA programme and once trained over 2,500 people on project-management techniques for the Canadian military. He is not someone his colleagues would associate with missed milestones. However, even experienced project managers sometimes find it hard to realize initially what could be a showstopper. Spanitz explains: 'I was managing a project for a medium-sized paper company here in the Chicago area.' The company had around 300 employees and was split across two locations. Spanitz was working with the IT programmers. 'Now, by saying I was "managing" the project,' he continues, 'in hindsight I would say I was cracking the whip, being more of a manager rather than a project manager.'

The project deadlines had been set by senior management without the involvement of the IT department and without a full understanding of the complexity of the project. 'Getting a project done "by Christmas" just had a nice ring to it,' Spanitz says. 'There was no consideration that I knew of about how realistic this deadline was.' The deadline was particularly tough, as between 80 and 90 per cent of the paper company's orders happened during the Christmas season. This put additional pressure on the IT department, which had to cope with business-as-usual problem solving and fire fighting. 'The programmers were all working 14- to 16-hour days, with pretty much everyone coming in on the weekends, myself included, just to even attempt to keep up with the schedule,' Spanitz, now president of the management consultancy Synergest, says.

One Friday, Spanitz came into the office at 8 a.m., late by his normal standards. He had been to see his doctor and was becoming increasingly aware of the ill-health spreading throughout his team. 'I walked by the printer, only to see the résumé of our lead programmer printing out at least 100 copies,' he says. 'On my way to my office, I passed another programmer, a very proper gentleman in his mid-fifties, with his head on his desk sobbing – no, more like a low wail. Being the emotion-avoider that I was, I ducked into my office to avoid getting pulled into some messy emotional issue. I looked at the clock and realized I had to do something and at this point really had nothing to lose.'

Spanitz realized that he had to take drastic action or he would end up losing his entire team. The deadline was completely unachievable, and the IT department was falling under a black cloud by trying to do the best they could. 'I loosened my tie, messed up my hair, took a couple of deep breaths and walked down to the conference room, where the senior management was having their weekly meeting,' Spanitz says. 'I walked into the meeting high on adrenalin and just started yelling "You guys are killing us back there. The whole department is sick and getting ready to quit. To Hell with your unrealistic deadlines! I'm giving everybody today off and we're not coming in this weekend!" I stormed out of the conference room still seeing red and made a beeline to the IT department. I quickly whispered loudly to the programmers to grab their stuff and get out – not to come in this weekend, and we'll figure stuff out on Monday. It was like an airplane evacuation as we all ran out the door.'

However, Spanitz did work that weekend. He used the time to put together a rough project plan for the work, which detailed how it could be completed in a reasonable timeframe. His calculations showed that the project could be delivered by the following April. 'I figured even if I was fired, I would still give it to my boss to show why I knew the deadlines were unreasonable,' Spanitz explains. 'On Monday, I went into the vice president of operation's office, who was my boss, and handed him the schedule.' Spantiz didn't know how to read his manager's behaviour. 'Then he said, "Oh good, I thought you were quitting" and looked at the schedule. He said: "You put on quite a performance – we didn't expect that from you." Before I could say anything, he said: "I think we need to include you in our planning meetings, so you can tell us, with your inside voice, how realistic our proposed deadlines are."'

Spanitz came to the realization that morning that the unrealistic deadline and the decreasing morale of his team was a showstopper. 'If the project manager does not speak up for his or her people, nobody will. Part of a project manager's job is to be an advocate and liaison between the project team and the executives,' he says. Monitoring his team's behaviour allowed him to identify just how low things were getting. 'Extended periods of overtime are counterproductive,' Spanitz adds. 'At that point, morale is destroyed, people will get sick, people will quit and productivity plummets. Sitting in front of a computer for longer time periods does not equal more work getting done.'

Since that impromptu meeting with the senior management team, Spanitz has found that on new projects, executives rarely try to force an unreasonable deadline, as long as he can explain why the deadline is unreasonable.

'Sometimes, theatrics are necessary to emphasize a point,' Spanitz concludes. 'I have never done a similar "performance" again, yet whenever I think back to that situation, I have no doubt that what I did was necessary and appropriate. The senior management team included me as the project manager in all future planning sessions involving the IT department. They listened to our discussion about project schedule feasibility, and over the next six months they reworked the pay structure for the programmers to increase pay and to introduce pay for weekend work. I am almost embarrassed to admit that sometimes I wonder if I should have performed sooner.'

You cannot work to resolve or get round the showstopper without first understanding exactly what it is. 'This involves investigation work to determine the dimensions and extent of the problem,' says Richard Murch in his book *Project Management: Best Practices for IT Professionals.*[85] 'Broadly, we define a problem as a deviation from an expected level of performance whose cause is unknown.' Getting to the bottom of the cause is essential, so call a crisis meeting with your team to sit down and work out why the situation has happened and what exactly has gone wrong. There will be a huge temptation to jump in immediately and suggest solutions and a plan of action for what to do next. However, bring your facilitation skills into play and make sure that everybody has the same understanding of the situation and its result before you allow the discussion to move on to analysing options for action.

HINT

Don't be tempted to simply accept the first plan of action. With a little bit of time, you and the team may well be able to come up with several alternatives. However, of course, in some situations there will be only one alternative: stop the project. In any situation where it is possible to continue the project, stopping it should also be analysed along with the other possible alternatives. This level of consideration is necessary because you and your project team need to be 100 per cent behind the decision. You need to understand why you are implementing this showstopper management plan and, more importantly, be able to explain it to your project board. As the project manager, you may have to take a bullish attitude to move the plan into action as, by their very nature, showstoppers don't often give you a lot of time to sit around and deliberate. Some of the most rewarding moments of project management can be in response to this kind of fire fighting.

GOLDEN RULES

You will know when a showstopper threatens the success of your project. Keep a calm head, analyse the problem and come up with a robust and effective management plan.

52 Learn how to facilitate

Facilitation is a skill that all project managers – all managers, for that matter – can benefit from developing. It is a method of participating in a group discussion and working towards the goal of getting the best out of the participants, in terms of both their behaviour and their contribution to the subject up for discussion.

a

MUSEUM MEETINGS

Naida Kendrick Culshaw was working at a medium-sized science museum in California as their advertising and promotions manager during a time of great change for the museum. 'The museum had undergone a large transition from a small gallery space of 20,000 square feet to 100,000 square feet and a sharp increase in staff, who were all moving very quickly and, because of that, occasionally out of sync with each other,' she says. 'I approached my manager and said that to do my job efficiently, I needed to have a sense of who was doing what and when, as I was responsible for promoting a product – the museum – which I didn't develop or control for the most part.'

Kendrick Culshaw proposed holding a cross-departmental meeting as the start of a project team, with the aim of bringing together around 15 key personnel to launch new exhibitions in a cohesive way. 'My facilitating skills then came in handy, as I had to manage the individual agendas and concerns of each member of the team,' Kendrick Culshaw explains. 'There was a feeling of needing to defend one's turf, and I wanted to bridge the communication gap and engage each individual to see how they could inspire each other to develop solutions to issues that may seem cut and dry.'

Kendrick Culshaw believes that as a facilitator, her role working with this team was to keep the group task-focused, to encourage creative thinking, to build consensus and to keep everyone involved. 'Not everyone approached this as a positive thing – far from it,' she adds. 'But I knew how to manage them – discussing issues offline, helping them see the variety of viewpoints that they may not agree with but are still valid and should be respected. I was the "neutral" person in the room, allowing a free flow of discussion while keeping things on track.'

This approach paid off, as Kendrick Culshaw is quick to explain. 'After we started the process and the members of the team could see the benefits of opening up and sharing their ideas and thoughts, we developed a dynamic group that were there to support each other, not to create roadblocks or extra hoops, either on purpose or just by accident.' The first exhibition launch on which the team worked together was a great success. 'The team pulled off an amazing exhibition launch, but in the process I discovered that I had helped to clear long-held misunderstandings by shedding light on the processes of each group and how those impact each other,' Kendrick Culshaw says. 'This was the start of a lovely relationship, and in the end I was asked to be the launch coordinator for all exhibitions and films for the museum.'

The cross-functional team may have ended up working together smoothly, but along the route there was still a lot of hard work for Kendrick Culshaw. 'For me, the energy spent before, during and after each meeting was listening and watching for nuances, content, body language and other feedback that was being shared by each team member,' she explains. 'By picking up on those signs, I helped facilitate the continued exchange and communication in between meetings. I had learnt from a workshop I attended that in a meeting, or just when speaking with a group, to be aware of two communication levels simultaneously: content – what's being discussed or decided – and process – how the group is functioning.' Kendrick Culshaw learnt to blend assertiveness with tact, discipline and humour through the process as well and became more aware of when to effectively intervene when the meeting veered off the subject.

Kendrick Culshaw believes that being a good facilitator is partly innate but partly a learnt skill. 'I really think I sort of "have" this ability,' she says. It has proven a useful skill across her career. After leaving the museum, Kendrick Culshaw moved to France and became president and vice-president of programmes of WICE, a non-profit cultural and educational association in Paris, where she has used her ability to recruit and motivate a vast team of staff and volunteers. 'It's something I think I honed from childhood. I was the oldest of three, so facilitating the childhood scrapes and arguments was my job, one which I managed with a diplomatic air, according to my mother. My parents always taught us how to look at all points of view, to put ourselves in someone else's shoes, then make comments on a situation.' She points out, however, that facilitation skills can be learnt on a certain level. 'It takes knowledge and talent to read between the lines, to feel a shift in the emotional level in a room and gauge the non-spoken signals that can make or break a session,' she says. 'The ability to "hear" what is being said and to then convey that message in a way that the listener "hears" what the key message is – even through their own filter. Those things can be learnt, but I feel that those who gravitate towards such a role have a bit of natural talent to build upon.'

Facilitation will not help resolve all situations, according to Kendrick Culshaw. 'I think that when you try to facilitate a situation where the parties are just set on their view and will not budge or open up, then facilitating may help them see other points of view but won't solve the issue.' She has some advice for people beginning to use facilitation skills in projects: 'Stay on task, have an agenda but let the meeting breathe. That means having some space and time to allow for free flow of ideas and acknowledge each as a contribution to the discussion. Then feed back what you've understood from the whole and define the next steps. Identify the goals to be accomplished and make sure that everyone understands what those goals are. It's amazing how many meetings you can attend and never really know why the meeting was called and what came out of it.'

'Facilitation is a way of providing leadership without taking the reins,' writes Ingrid Bens in her book *Facilitating with Ease!* 'Most important, you help [group] members define and reach their goals.'[86] Facilitation used to be linked with the set of skills needed to be a good trainer, and the 1980s saw facilitation moving out of the classroom with the advent of total quality management (TQM).[87] These days, it is well recognized that facilitation skills will help you run project meetings and workshops, but 'facilitation' itself is hard to define precisely and you probably do it without being aware of it in many of your meetings already. Try to become more conscious of how you behave in meetings, which will help you identify which areas you need to improve. For example, do you:

- always make sure everyone contributes?
- create an environment where it feels safe to share ideas?
- actively listen to each individual?
- summarize important points and reflect them back to the group?
- ask lots of questions?
- tailor your style depending on the audience and task in hand?
- plan your meetings in advance?
- understand group dynamics?
- have experience of how groups resolve issues in a workshop environment?

If you answered no to any of those questions, then brushing up your facilitation skills would probably be a good idea. However, you can always bring in an external facilitator – a consultant, a more experienced project manager or someone from your HR or training department – to help out for meetings that you expect to be particularly tricky to handle. You can offer the same service to other project or operational managers to build up your exposure to these situations.

Judith Kolb and William Rothwell from Pennsylvania State University asked 63 expert facilitators from the International Society for Performance Improvement, an association dedicated to improving productivity and performance in the workplace, about what makes a good facilitator. The responses to their questionnaire showed that the top competencies deemed important for small-group facilitators were:

- active listening;
- skilful handling of questions;
- awareness of group dynamics;
- ability to paraphrase contributions from the group;
- being able to animate the group and stimulate creativity.[88]

Their study also concluded that there are useful practical techniques for successful facilitation, and a facilitator should have more than one tucked away: 'There are myriad decision-making/problem-solving techniques available to help groups manage information and reach decisions,' they write. 'An experienced facilitator should be knowledgeable in a variety of methods so that s/he is not "forcing" a technique that does not fit the situation.'[89]

Tools and techniques will get you only so far. 'Outstanding facilitation is, of course, much more than a smooth presentation style or being adept with flip charts or visual aids,' writes John van Maurik in an article for *Leadership and Organization Development Journal*. 'It is about achieving change, enabling excellence, or empowering groups to achieve results for themselves.'[90] In a project context, the results may be a clear understanding of project scope or requirements, an appreciation of what went wrong and what can be done to fix it, the understanding of what other team members do and how they can support each other or a whole host of other objectives. It will be up to you to decide when 'heavy' facilitation is required – with flipcharts, sticky notes, warm-up games (icebreakers) and so on – and when you can get away with facilitating with a lighter touch.

GOLDEN RULES

Some meetings need more than just an agenda and someone to chair. Learning to facilitate can help you get the best out of your time and your team.

53 Get a mentor

Having a mentor can be advantageous for your project-management career. Mentors offer their skills, experience and, critically, networks to their protégés or mentees. You can use a mentor to test ideas in a safe environment, ask for advice or just sound off about a particular problem.

ANECDOTE

When Marina Sampson, a customer service professional, stepped away from running a successful team and into project management, she realized that the skills required were quite different. She found an experienced project manager whom she felt would be a good mentor. Her mentor worked with her to help transform her experience of dealing with customers into a new toolkit for dealing with stakeholders and project teams over which she had no line-management authority. 'I have learnt so much from my mentor,' Sampson says. 'Even though we work in very different styles, she will look for the best result for me.' She arranged informal monthly meetings with her mentor for a catch-up over coffee and more structured sessions around the technical side of project management if and when they were needed. 'It's important to find someone you can be honest with,' Sampson advises. 'Whatever the issue is, we can work together to figure out a resolution.'

Kimberly McDowall-Long surveyed the available research into the success of mentoring relationships in 2004 and concluded that it can have a very beneficial effect on the protégé's career in terms of improved job satisfaction, more rapid promotion, higher salaries and increased access to organizational key players.[91] However, choosing a mentor should not be rushed into. If your company has a formal mentoring programme, you may be allocated a senior manager to support you in the role of mentor. You can still have a mentor even if your company does not have such a scheme, although you will have to seek out and approach them yourself. Either way, hopefully the person chosen as your mentor will play a supportive role in your long-term future.

McDowall-Long identified two groups of characteristics displayed by effective mentors: interpersonal skills and technical expertise. When you begin to think about who you would like as a mentor, consider their competence in both those areas. If you decide to use a formal company scheme ensure you will have the opportunity to swap mentors if the two of you do not click; it can be quite a personal relationship, so it is important you get on. Give some thought as well to perhaps having different mentors for different things. The technically brilliant budget whiz may help you out of some awkward financial moments, but she may not be the one with whom to discuss sensitive communications issues.

When drawing up your shortlist of potential managers to approach as your mentor, consider the following:

- What sort of person do you feel comfortable opening up to?
- How much more senior than you do you want your mentor to be? Perhaps seniority is not as important as their project management experience or their character.
- Would you prefer a male or female mentor? Of what age?
- Do you want someone with an established network who could perhaps help you achieve your career goals?
- Watching them in their interactions with others, do they have the communication skills to be able to give you constructive feedback and set you challenging targets, while remaining supportive?

Once you have selected possible candidates, find a convenient time to approach them one by one. Explain to your top choice what you expect of a mentor. How many meetings per month would you want to arrange? Would they be on a regular basis or ad hoc as you need them? How often do you expect to call on them, and for what? Once you have outlined your objectives and explained what your prospective mentor is getting into, ask them to consider it and let you know in due course whether they are prepared to step into the role. Being asked to be someone's mentor is flattering, but it is also a big commitment – informal mentor–protégé relationships can last up to six years.[92] Let your candidate think about it, and don't take it personally if they say no. You might have cause to work with them in the future, so don't feel rejected if they turn you down – and, of course, maintain a professional relationship. It is unlikely that their decision was based on working with you personally. If they do say no, move on to the second candidate on your list and ask them. Alternatively, ask your top choice politely if they can suggest someone else: they obviously move in the right circles or you would not have selected them, so they may well know of someone outside your personal circle of acquaintance who would be a suitable mentor. And they will probably be pleased that you value their opinion enough to ask.

HINT

If the pool of possible candidates within your company has been exhausted, look elsewhere. There are benefits to having a mentor from your own company (for example, increased internal recognition, improved promotion prospects), but there are also advantages to having someone completely independent.

There are professional networking groups for all professions, so you should be able to find someone somewhere doing a similar thing to you.

GOLDEN RULES

Mentoring can provide you with a great career advantage, access to another set of contacts and a hotline to technical project management expertise, but you need to choose your mentor carefully in line with your own expectations of how the relationship will develop.

54 Do documentation

Documentation is one of the onerous jobs of a project manager: compiling initiation documents, plans, product descriptions, roles and responsibilities and the other countless pieces of paperwork that form part of your project dossier. However, documentation is essential, as it helps turn the nebulous ideas of your project sponsor into a fully fledged, well-understood, tangible project.

ANECDOTE

Adam, a telecoms engineer based in Cardiff, remembers his first project well. 'I had just started with my company,' he says. 'I was managing the implementation of a new phone system for a reasonably small client. We had a couple of meetings and it was going well.' Six weeks later, the clients started to ask questions. 'I couldn't remember whether we had agreed to weekly meetings or whether they had just happened. The client decided it was too frequent and pulled their guys out. And because the senior manager there didn't know what the sponsor should do, when I asked him to intervene he said he had never agreed that it was part of his role on the project and he wouldn't step up.' Adam learned the hard way that project documents have a useful role as memory joggers. Fortunately, his project team managed to adapt to fortnightly meetings and the telephone system was implemented on time.

Project documentation serves three functions: clarification, concentration and confirmation.

- **Clarification**: the act of writing things down allows you to clarify vague ideas. It guarantees that everyone involved in the project has a clear understanding of the aims and objectives, scope, plan and assumptions. The act of creating these documents with input from critical stakeholders facilitates the discussion of what everyone is expecting from the project. Any vagaries or multiple interpretations can be cleared up at this point.
- **Concentration**: a roles and responsibilities statement sets out exactly what is expected of each member of the project team, at every level. It ensures people know what they're signing up for and is especially useful as a starting point for discussion with the project sponsor about their role, as senior managers are often not clear about what sponsoring a project entails. Terry Cooke-Davies studied 136 projects across 23 organizations over six years and concluded that adequate documentation of project responsibilities was one of the 12 factors that make

a project successful.[93] It also correlates statistically to completing a project on time. Documentation encourages people to concentrate in another way too – the review and signoff process ensures you have a physical agreement in the form of a signature from all key decision makers. This is tantamount to an informal internal contract. Although managers do sign off project documents without reading them, asking them to sign their name does increase the pressure. It will never guarantee complete buy-in to the document, but it is a step in the right direction. To check whether her documentation was read, one project manager sent out a project-initiation document that listed one of the responsibilities of her sponsor as sending her chocolate on a weekly basis. The sponsor noticed it, but none of the other document reviewers mentioned it.

- **Confirmation**: while a project exists only in people's heads, it will struggle to be taken seriously. Sarah Blackburn, in her paper on project networks, explains that 'the project also has to create itself and maintain itself for the time needed to effect delivery . . . [It] craves embodiment: a code name, a project room, a logo, t-shirts – these are peripherals – the core is the project documentation.'[94] Documentation makes the project 'real'. It moves the project from the realm of a good idea into a realm where a professional project manager can turn it into a tangible activity and delivery.

For your project, consider what documents you will need to produce. As a minimum, you will need those covered in Table 54.1. Your organization might expect you to produce a quality plan or other documentation too.

Once a document is written, it should be circulated for review by anyone who will be impacted by the content. Give your reviewers a deadline by which to send you their comments – add that if you hear nothing, you will assume that they have no comments to make. Include the feedback you have received; if it has changed the document substantially, reissue it for review. An example of a substantial change would be if a reviewer from marketing added in additional work for sales to do. Reviewers should comment in detail on parts that impact their own teams, but their comments about the involvement of other teams should always be discussed with someone from those teams. When all the substantial changes have been validated by the right people, the document is ready to be signed. In your covering memo or email, make a note of any minor changes you have made that the signatory has not yet seen. Sending a copy by email for the signatories' reference is a good idea. It also counteracts any wariness they have when you present them with a hard copy of the same document.

Circulate one hard copy for signature. Everyone should sign the same copy of the document, so it will end up looking a bit tatty by the time you get it back. The signature area should appear in the document as in Table 54.2.

TABLE 54.1 *Standard project documents*

Document	Function
Project-initiation document (PID)	The PID comes in many forms and is the first project document produced after approval is given for the project to start. It covers project scope, objectives, the principles and methodology under which the work will be carried out, and high-level budget. It may incorporate high-level plan details and initial risk and issue register
Plan	A plan can be made up of subdocuments, e.g. detailed description of all the deliverables. As a minimum, a plan is a list of the tasks to be done, plus the dates they will start and finish and who is to do them
Roles and responsibilities statement	Explains the roles within the project (sponsor, project manager, business owner, IT team leader and so on) along with the responsibilities that role carries. It should be as specific as possible
Initial risk and issue register	A living document that will be added to as you go through the project and continue to manage existing risks and issues and add new ones. At the beginning of the project, this document sets out what it is you know already about the risks and issues associated with the project
Requirements document	Details exactly what any new process, product or system should do and is put together based on the requirements of the customer or business
Post-project documentation	At the end of the project, you need to get approval for the project to be closed down. This is normally in the form of a post-project document, which follows a post-project review meeting. A shorter document, a project close-down document, can be used instead for formal authority to stop the project, especially if the project did not reach its planned end

TABLE 54.2 *Example of a signature area on a typical project document*

The signatories below authorize the work detailed in this document to be carried out

Name	Role	Signature	Date
Meera Rantasha	Project sponsor, director of operations		
Simon Wilkinson	Project manager		
Natasha Culshaw	Head of sales		

HINT

If you are seeing several signatories together at, for example, a project board meeting, take the document along for signoff then. Alternatively, use the internal mail. Send it to your first signatory, who should be your sponsor, and ask for it to be returned to you. It is too complicated to ask managers to pass it on to the next person on the list. Too many documents get lost this way. It also gives you the opportunity to make a photocopy of the signature page before sending it out again in case the director's filing system or the internal mail takes the whole document astray. Then, if the document is lost, you have to approach only the people who have not yet signed it, avoiding starting from the beginning again and bothering people who have signed once already.

Once you have a document signed by everyone, including you, type the names and dates into an electronic copy and send everybody the approved, final version as a record of what has been agreed. See Chapter 17 for more on version control for documents.

GOLDEN RULES

By producing documentation, you are giving the project shape and structure, helping to generate buy-in and ownership and also ironing out any vagaries early on.

55 Don't be afraid to suggest they pull the plug

The average large company, running around 150 projects at any one time, loses £13 million a year by not stopping projects that are failing.[95] Suggesting your project is scrapped can be a difficult message to give your sponsor and stakeholders. However, the project manager's role is partly to direct the work and partly to provide an objective position on how the work is done – and if that means suggesting stopping everything and starting again, or even not starting again, then that is part of your role too.

a

SECURING A GOOD RESULT

A junior manager working in the security industry was called in to help with the communications on a particularly sensitive project. A new piece of legislation meant all security guards had to be licensed, a process that cost around £200 per employee. The project team had been working on applying for licences within the deadline and the payment process to pass the cost on to the individual guards. The manager recognized the human implications of passing a £200-bill on to staff who, in the main, were receiving the minimum wage. 'The impact hadn't been thought through,' he says. When he understood the project's key stakeholders fully, he immediately approached the right people and recommended the initiative was stopped. The alternative proposal was for each of the security company's clients to meet the bill for licensing instead of passing the charge on to the employee. The project was restarted with this as the end goal, saving the company an embarrassing situation with unions and staff.

By continuing with something you know is doomed, you risk being personally tarnished with the project's failure and being criticized for not making your superiors aware of the situation. Your sponsor may well insist on the work being completed, even once you have presented the most compelling of arguments for why that is not a practical conclusion. However, the important point is that you will have done your bit in flagging the position and will, of course, maintain your risk and issue log regularly to make certain that your concerns are noted and action is being taken to mitigate the risk of failure.

Keeping note of those risks and issues during a project that is on the brink of failure is particularly important. If your sponsor will not listen to your advice to cut the project's losses and get out now, then you can at least use the risk and issue log to explain the consequences of completion. Sometimes that alone can have a sobering effect. There is nearly always another way to achieve similar objectives – so if you can think of another solution, point it out.

There are sensitivities to take into account when pointing out that a project is never going to complete successfully. In some cultures, regardless of what you believe, having that discussion with your project sponsor would be unthinkable. 'Failure, operationalized as a cancelled project or a bankrupt company, is not as grave in Silicon Valley as it would be in Taipei, where success and failure are bound up with *mianzi*, or "face", a concept that is culturally specific,' conclude three anthropologists in their research about work relationships.[96]

GOLDEN RULES

Do not be afraid to challenge senior people. Remember that not all projects are started from a basis of a well-thought-through and competent idea. If the project is going to be a failure and you know it, explain why it should be stopped.

56 Archive effectively

It is the end of your project: the sponsor thinks you did an excellent job and the people who have to live with the change are happy. But it's not quite over. All those documents, minutes, decisions, test notes and emails that you have accumulated over the length of the project have to go somewhere and it would not be right to dump them in the recycling bin. This is where archiving comes in. At some point in the future, someone is bound to ask you why a particular task was or was not done. Archiving will mean you can get hold of the information easily to answer their question, but the files won't be cluttering up your desk while you are trying to work on something else. Having the files available will also help other project managers who find themselves doing a similar project in the future.

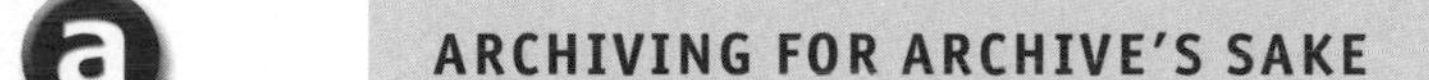

ARCHIVING FOR ARCHIVE'S SAKE

John, a journalist, was updating the archives of the regional southern England weekly newspaper where he was working when he came across a file called 'Bubbles'. 'I wasn't sure why the paper would have a whole archive folder on the subject of bubbles – we used the archives as a reference for future news stories,' he says. When John looked, there was one clipping in the folder. The story was about a family reunion with the headline 'Daughter bubbles with joy'.

HINT

As a rule of thumb, keep your files easily accessible for one year, perhaps in a team cupboard or held by the department that now runs the business-as-usual activity. Then archive them.

Before tidying up your files to put them all in one place, check any regulatory requirements about storing different types of data and the length of time records are normally kept in your company. Print out any relevant emails lurking on your hard drive or burn your email archives to CD. Label each folder clearly with a contents list. Then label each box with the project name. Number folders (one of three, two of three etc.) and boxes so if they need to be retrieved from off-site storage you can be sure you have them all.

HINT

Make your labels transparent and logical. As your project files may be used by someone else in the future, don't be cryptic.

It might even be you in a few years who requires access to information, and what was crystal clear then may have faded in your memory with time.

GOLDEN RULES

A project isn't over when the post-project review is signed. Archive your project data to be sure the full history is available if it is ever needed again.

Appendix

RISK LOG

Risk ID	Title	Date raised	Owner	Impact	Likeli-hood	Status	Notes and actions

Impact categories

Minor: impact of less than £40,000, or no reputational damage.
Moderate: impact of between £40,000 and £200,000, or possible small amount of reputational damage.
Significant: impact of between £200,000 and £1 million, or moderate amount of reputational damage.
Severe: impact of over £1 million, or destructive reputational damage.

Likelihood categories

Remote: less than 1 in 1,000 chance of occurring – not foreseeable within 5 years.
Unlikely: less than 1 in 100 chance of occurring – could happen within 5 years.
Possible: less than 1 in 10 chance of occurring – could happen within a year.
Probable: more than 1 in 10 chance of occurring – imminent.

ISSUE LOG

Issue ID	Title	Date raised	Owner	Prior-ity	Status	Notes and actions

Priorities

High, medium, low.

CHANGE LOG

Change ID	Description	Date raised	Owner	Related documents	Priority	Date impact assessment completed	Outcome

Priorities

Critical, important, cosmetic, optional.

Outcome categories

Reject, accept, pend/postpone.

Notes

SECTION 1

1. Sauer, C. and Cuthbertson, C. (2004) *The state of IT project management in the UK 2002–2003.* www.computerweeklyms.com/pmsurveyresults/surveyresults.pdf (accessed 5 March 2006).
2. Doss, G. M. (2005) *IS Project Management Handbook.* Aspen, New York.
3. National Audit Office (2000). *The Millennium Dome.* Report by the Comptroller and Auditor General. HC936 Session 1999–2000, 9 November 2000. London: The Stationery Office.
4. Shim, J. K and Siegel, J. G. (2005) *Budgeting Basics and Beyond.* Wiley, New York.
5. Keil, M., Mann, J. and Rai, A. (2000) Why software projects escalate: an empirical analysis and test of four theoretical models. *MIS Quarterly,* 19, 421–447.
6. Flyvberg, B., Holm, M. K. and Buhl, S. L. (2002) Understanding costs in public works projects: error or lie? *Journal of American Planning Association,* 68, 279–295.
7. Eden, C., Ackermann, F. and Williams, T. (2005) The amoebic growth of project costs. *Project Management Journal,* 36, 15–27.
8. Portny, S. E. (2001) *Project Management for Dummies.* Wiley, New York. [Appendix B has an excellent introduction to earned value analysis, including easy-to-follow worked examples.]
9. For more on this project, see GAO-04-611 (2004) *Nuclear waste: absence of key management reforms on Hanford's cleanup project adds to challenges of achieving cost and schedule goals.* US General Accounting Office, Washington, DC.
10. GAO-05-123 (2005) *DOE's management of major projects.* US General Accounting Office, Washington, DC.
11. Schulte, R. (2002) *Modern cost management.* Welcom white paper. www.welcom.com/content.cfm?page=704 (accessed 21 April 2006).
12. Wright, J. N. (1997) Time and budget: the twin imperatives of a project sponsor. *International Journal of Project Management,* 15, 184.
13. Heldman, K. (2003) *Project Management JumpStart.* Sybex, San Francisco.
14. ibid.
15. Research commissioned by Microsoft and Corporate Project Solutions. Cited in Lane, K. (2004) Worrying flaws in PM practice. *Project Manager Today,* April, 4.
16. Cicmil, S. J. K. (1997) Critical factors of effective project management. *TQM Magazine,* 9, 392.

17. Multiplex released details of the Wembley National Stadium project and its impact of the FY2005 in a press release on 30 May 2005, available online at www.multiplex.com.au/page.asp?partid=294&ID=170 (accessed 10 March 2006).
18. Bradbary, D. and Garrett, D. (2005) *Herding Chickens: Innovative techniques for project management.* Jossey-Bass, San Francisco, p. 27.
19. op. cit., p. 28.
20. White, D. and Fortune, J. (2002) Current practice in project management: an empirical study. *International Journal of Project Management*, 20, 1–11.

SECTION 2

21. Sauer, C. and Cuthbertson, C. (2004) *The state of IT project management in the UK 2002–2003.* www.computerweeklyms.com/pmsurveyresults/surveyresults.pdf (accessed 5 March 2006).
22. This six-step approach has been adapted from the model presented in Everett, C. (2005) How to ensure pilot projects are successful. *Computing*, 3 November, 48–52.
23. Duncan, B. (2003) *Ignorance is risk.* www.projectsatwork.com/article.cfm?ID=217483 (accessed 12 January 2006).
24. Cicmil, S. J. K. (1997) Critical factors of effective project management. *TQM Magazine*, 9, 394.
25. Crawford, L. (2005) Senior management perceptions of project management competence. *International Journal of Project Management*, 23, 7–16.
26. Cooke-Davies, T. (2002) The 'real' success factors on projects. *International Journal of Project Management*, 20, 188.
27. White, T. (2004) *What Business Really Wants from IT: A collaborative guide for business directors and CIOs.* Butterworth-Heinemann, Oxford.
28. GAO-05-711 (2005) *Foreign assistance: Middle East Partnership Initiative offers tools for supporting reform, but project monitoring needs improvement.* US Government Accountability Office, Washington, DC.
29. Fink, D. (2003) Case analyses of the '3 Rs' of information technology benefit management: realise, retrofit and review. *Benchmarking*, 10, 367–381.
30. Le Guin, U. K. (1998) *Steering the Craft.* Eighth Mountain Press, Portland, OR.
31. Raynes, M. (2002) Document management: is the time now right? *Work Study*, 51, 303–308.
32. This case study appears in Skinner, D. (2004) Evaluation and change management: rhetoric and reality. *Human Resource Management Journal*, 14, 519.
33. The author acknowledges Martin Schindler and Martin J. Eppler for the term 'project amnesia' in the paper Schindler, M. and Eppler, M. J. (2003)

Harvesting project knowledge: a review of project learning methods and success factors. *International Journal of Project Management*, 21, 219–228.

34. HC 742 (2005) *Achieving value for money in the delivery of public services.* House of Commons Committee of Public Accounts, The Stationery Office, London.
35. HC1159-11 (2004) *Ministry of Defence Major Projects Report 2004: project summary sheets. Report by Comptroller and Auditor General.* The Stationery Office, London.
36. HC1159-1 (2004) *Ministry of Defence Major Projects Report 2004. Report by Comptroller and Auditor General.* The Stationery Office, London.
37. Elkington, P. and Smallman, C. (2002) Managing project risks: a case study from the utilities sector. *International Journal of Project Management*, 20, 56–57.
38. Risks do not have to be limited to negative outcomes, although this is the traditional way of defining project risk. For an analysis of how to extend the risk-management process to include the management of positive risk, see Hillson, D. (2002) Extending the risk process to manage opportunities. *International Journal of Project Management*, 20, 235–240.
39. Baccarini, D., Salm, G. and Love, P. E. D. (2004) Management of risks in information technology projects. *Industrial Management and Data Systems*, 104, 289.
40. For more information about risk registers and how risks can be logged in practice, see Patterson, F. D. and Neailey, K. (2002) A risk register database system to aid the management of project risk. *International Journal of Project Management*, 20, 365–374. This paper provides an overview of a risk database used in an automotive company and is a useful study into how risk registers can be made to work.
41. Cooke-Davies, T. (2002) The 'real' success factors on projects. *International Journal of Project Management*, 20, 186.
42. Elkington, P. and Smallman, C. (2002) Managing project risks: a case study from the utilities sector. *International Journal of Project Management*, 20, 55.
43. Baccarini, D., Salm, G. and Love, P. E. D. (2004) Management of risks in information technology projects. *Industrial Management and Data Systems* 104, 291.
44. See, for example, Khalfan, A. (2003) A case analysis of business process outsourcing project failure and implementation problems in a large organization of a developing nation. *Business Process Management Journal*, 9, 745–759; Dvir, D., Raz T. and Shenhar, A. J. (2003) An empirical analysis of the relationship between project planning and project success. *International Journal of Project Management*, 21, 89–95; Turner, J. R. (2004) Five necessary conditions for project success. *International Journal of Project Management*, 22, 349–350.

45. I am indebted to Neville Turbit for explaining this exercise to me.

SECTION 3

46. Goodman, J. and Truss, C. (2004) The medium and the message: communication effectively during a major change initiative. *Journal of Change Management*, 4, 225–226.
47. Obeng, E. (2003) *Perfect Projects*. Pentacle Works, Beaconsfield.
48. Trust in a workplace environment is discussed in an interesting study by English-Lueck, J. A., Darrah, C. N. and Saveri, A. (2002) Trusting strangers: work relationships in four high-tech communities. *Information, Communication and Society*, 5, 90–108.
49. Cited in Thomas, D. (2004) Project of the Year Awards: Nectar: Nectar improves customer service with web overhaul. www.computing.co.uk/computing/analysis/2075963/project-awards-nectar (accessed 25 July 2006).
50. White, D. and Fortune, J. (2002) Current practice in project management: an empirical study. *International Journal of Project Management*, 20, 1–11.
51. Trompenaars, F. and Woolliams, P. (2003) A new framework for managing change across cultures. *Journal of Change Management*, 3, 361–375.
52. I gratefully acknowledge Barry Shore and Benjamin J. Cross for posing this question in their paper Shore, B. and Cross, B. J. (2005) Exploring the role of national culture in the management of large-scale international science projects. *International Journal of Project Management*, 23, 55–64.
53. CERFDOE Final Report 071204 (2004) *Independent research assessment of project management factors affecting department of energy project success*. Civil Engineering Research Foundation, Reston, VA, p. 18.
54. GAO-01-459 (2001) *Computer-based patient records: better planning and oversight by VA, DOD and IHS would enhance health data sharing*. US Government Accountability Office, Washington, DC.
55. GAO-02-703 (2002) *Veterans Affairs: sustained management attention is key to achieving information technology results*. US Government Accountability Office, Washington, DC.
56. GAO-04-687 (2004) *Computer-based patient records: VA and DOD efforts to exchange health data could benefit from improved planning and project management*. US Government Accountability Office, Washington, DC.
57. GAO-07-755T (2005) *Systematic data sharing would help expedite service members transition to VA Services*. US Government Accountability Office, Washington, DC.

58. White, D. and Fortune, J. (2002) Current practice in project management: an empirical study. *International Journal of Project Management*, 20, 1–11.
59. Hacker, M. (2000) The impact of top performers on project teams. *Team Performance Management*, 6, 88.
60. If you are lucky enough to get to choose your own team, Meredith Belbin's (www.belbin.com) research into team roles might be interesting.
61. George Cowie, a senior consultant at MaST, presents this case study in his paper Cowie, G. (2003) The importance of people skills for project managers. *Industrial and Commercial Training*, 35, 256–258.
62. Obeng, E. (2003) *Perfect Projects*. Pentacle Works, Beaconsfield, p. 107.
63. For more on this case study, see Tibbitts, S. (2005) *The morning meeting*. http://appl.nasa.gov/ask/issues/20/features/20_morning_tibbetts.html (accessed 17 September 2005).
64. Boddy, D. and Paton, R. (2004) Responding to competing narratives: lessons for project managers. *International Journal of Project Management*, 22, 231.
65. Bourne, L. and Walker, D. H. T. (2005) Visualising and mapping stakeholder influence. *Management Decision*, 43, 650.

SECTION 4

66. Center for Business Practices (2003) *Project management: the state of the industry*. A summary of this research report is available online at www.pmsolutions.com/articles/pdfs/general/industry_news.pdf (accessed 3 March 2006).
67. Lissak, R. and Bailey, G. (2002) *A Thousand Tribes: How technology unites people in great companies*. Wiley, New York.
68. White, D. and Fortune, J. (2002) Current practice in project management: an empirical study. *International Journal of Project Management*, 20, 1–11.
69. Deckro, R. F. and Hebert, J. E. (2002) Modeling diminishing returns in project resource planning. *Computers and Industrial Engineering*, 44, 20. This paper presents a series of models for identifying the point of diminishing return.
70. Shrub sets out his project planning model for project segmentation in Shrub, A. (1997) Project segmentation: a tool for project management. *International Journal of Project Management*, 15, 15–19.
71. Reiss, G. (1995) *Project Management Demystified: Today's tools and techniques*. Spon Press, London.
72. For more on this case study, see Whittle, S. (2004) Stormy Weather. *Computer Weekly*, 3 August, 24–25.
73. Herroelen, W. and Leus, R. (2005) Project scheduling under uncertainty: survey and research potentials. *European Journal of Operational Research*, 165, 289.

74. Herroelen, W. and Leus, R. (2004) The construction of stable project baseline schedules. *European Journal of Operational Research*, 156, 550.
75. See, for example, Herroelen and Leus's 2004 article, which sets out a mathematical programming model for designing a baseline project schedule.
76. CERFDOE Final Report 071204 (2004) *Independent research assessment of project management factors affecting Department of Energy project success*. Civil Engineering Research Foundation, Reston, VA.
77. Finch, C. (2005) *To be accurate* . . . www.projectsatwork.com/content/Articles/226867.cfm (accessed 9 September 2005).
78. Research commissioned by Microsoft and Corporate Project Solutions, cited in Lane, K. (2004) Worrying flaws in PM practice. *Project Manager Today*, April, 4.
79. 3M survey (1998) cited in Simon, P. and Murray-Webster, R. (2005) Efficient and effective meetings: essential but elusive? *Project Manager Today*, June, 16.
80. For more on this case study, see Tibbitts, S. (2005) *The morning meeting*. http://appl.nasa.gov/ask/issues/20/features/20_morning_tibbetts.html (accessed 17 September 2005).
81. Garcia, A.C.B., Kunz, J. and Fischer, M. (2005) Voting on the agenda: the key to social efficient meetings. *International Journal of Project Management*, 23, 20.
82. Brown, M. and Draper, S.W. (1998) *Tips on running a video conference*. www.psy.gla.ac.uk/˜steve/resources/VidConfTips.html (accessed 17 September 2005).

SECTION 5

83. Greene, R. and Elffers J. (2000) *The 48 Laws of Power*. Profile Books, London.
84. Crawford, L. (2005) Senior management perceptions of project management competence. *International Journal of Project Management*, 23, 7–16.
85. Murch, R. (2001) *Project Management: Best Practice for IT Professionals*. Prentice Hall, Upper Saddle River, NJ.
86. Bens, I. (2005) *Facilitating with Ease!* Jossey-Bass, San Francisco.
87. Nelson, T. and McFadzean, E. (1998) Facilitating problem-solving groups: facilitator competences. *Leadership and Organization Development Journal*, 19, 73.
88. Kolb, J. A. and Rothwell, W. J. (2002) Competencies of small group facilitators: what practitioners view as important. *Journal of European Industrial Training*, 26, 201.
89. op. cit., 202.
90. van Maurik, J. (1994) Facilitating excellence: styles and processes of facilitation. *Leadership and Organization Development Journal*, 15, 34.

91. McDowall-Long, K. (2004) Mentoring relationships: implications for practitioners and suggestions for future research. *Human Resources Development International*, 7, 519–534.
92. Ragins, B. R. and Cotton, J. L. (1999) Mentor functions and outcomes: a comparison of men and women in formal and informal mentoring. *Journal of Applied Psychology*, 84, 529–549.
93. Cooke-Davies, T. (2002) The 'real' success factors on projects. *International Journal of Project Management*, 20, 185–190.
94. Blackburn, S. (2002) The project manager and the project-network. *International Journal of Project Management*, 20, 202.
95. Research commissioned by Microsoft, cited in *Computer Weekly*, 17 May 2004.
96. English-Lueck, J. A., Darrah, C. N. and Saveri, A. (2002) Trusting strangers: work relationships in four high-tech communities. *Information, Communication and Society*, 5, 95.

Further reading

SECTION 1

Chow, C. W., Harrison, P., Lindquist, T. and Wu, A. (1997) Escalating commitment to unprofitable projects: replication and cross-cultural extension. *Management Accounting Research, 8, 347–361.*

Elkjaer, M. (2000) Stochastic budget simulation. *International Journal of Project Management*, 18, 139–147.

Frow N., Marginson, D. and Ogden, S. (2005) Encouraging strategic behaviour while maintaining management control: multi-functional project teams, budgets, and the negotiation of shared accountabilities in contemporary enterprises. *Management Accounting Research*, 16, 269–292.

Powell, F. (2005) Has EVA come of age? *Project Manager Today*, December, 20–22.

Rachlin, R. (1999) *Total Business Budgeting: A step-by-step guide with forms.* Wiley, New York.

Rasmussen, N. and Eichorn, C. J. (2000) *Budgeting: Technology, trends, software selection and implementation.* Wiley, New York.

Ruchala, L. V., Hill, J. W. and Dalton, D. (1996) Escalation and the diffusion of responsibility: a commercial lending experiment. *Journal of Business Research*, 37, 15–26.

Webb, A. (2003) *Using Earned Value: A project manager's guide.* Gower Publishing, Aldershot.

SECTION 2

Bartlett, J. (2005) *Right First and Every Time.* Project Manager Today Publications, Hook.

Bing, L., Akintoye, A., Edwards, P.J. and Hardcastle, C. (2005) The allocation of risk in PPP/PFI construction projects in the UK. *International Journal of Project Management*, 23, 25–35.

Cooper, D. F., Grey, S., Raymond, G. and Walker P. (2004) *Project Risk Management Guidelines: Managing risk in large projects and complex procurements.* Wiley, San Francisco.

Eveleigh, C. (2005) Do post-project reviews really pay? *Project Manager Today*, November, 8–11.

Hameri, A. (1997) Project management in a long-term and global one-of-a-kind project. *International Journal of Project Management*, 15, 151–157.

Heldman, K. (2005) *Project Manager's Spotlight on Risk Management.* Sybex, San Francisco.

Oomens, M. J. H. and van den Bosch, F. A. J. (1999) Strategic issue management in major European-based companies. *Long Range Planning*, 32, 49–57.

Patterson, D. (2004) Improving project decision making and reducing exposure through risk management. Welcom white paper. www.welcom.com/content.cfm?page=708 (accessed 24 July 2006).

Van Well-Stam, D., Lindenaar, F., van Kinderen, S. and can den Bunt, B. P. (2004) *Project Risk Management: An essential tool for managing and controlling projects.* Kogan Page, London.

Webster, G. (1999) Project definition: the missing link. *Industrial and Commercial Training*, 31, 240–244.

Williams, T. (2004) Identifying the hard lessons from projects – easily. *International Journal of Project Management*, 22, 273–279.

SECTION 3

Bartlett, J. (2005) Expectation setting: fundamental to achieving project quality. *Project Manager Today*, June, 8–10.

Bradbury, A. (2000) *Successful Presentation Skills*, 2nd edn. Kogan Page, London.

Bridges, W. (2003) *Managing Transitions: Making the Most of Change*, 2nd edn. Da Capo Press, Cambridge.

Cook, S. (2001) Creating a high performance culture through effective feedback. *Training Journal*, August, 16–17.

Fox, C. (2005) Cross-border disconnection? *Project Manager Today*, June, 4–6.

Keller, G. (2004) *Think globally.* www.projectsatwork.com/articles/articlesPrint.cfm?ID=219168 (accessed 17 November 2005).

Maggio, R. (2005) *The Art of Talking to Anyone: Essential people skills for success in any situation.* McGraw-Hill, New York.

Turner, J. R. (ed.) (2003) *People in Project Management.* Gower, Aldershot.

SECTION 4

Calhoun, K. M., Deckro, R. F., Moore, J. T., Chrissis, J. W. and Van Hove, J. C. (2002) Planning and re-planning in project and production scheduling. *Omega*, 30, 155–170.

Devaux, S. (1999) *Total Project Control: A manager's guide to integrated project planning, measuring and tracking.* Wiley, New York.

Gardiner, P. D. and Ritchie, J. M. (1999) Project planning in a virtual world: information management metamorphosis or technology going too far? *International Journal of Information Management*, 19, 485–494.

SECTION 5

Greenly, D. and Carnall, C. (2001) Workshops as a technique for strategic change. *Journal of Change Management*, 2, 33–46.

Keegan, A. E. and Den Hartog, D. (2004) Transformational leadership in a project-based environment: a comparative study of the leadership styles of project managers and line managers. *International Journal of Project Management*, 22, 609–617.

Kirk, P. and Broussine, M. (2000) The politics of facilitation. *Journal of Workplace Learning: Employee Counselling Today*, 12, 13–22.

Laborde, G. Z. (2001) *Influencing with Integrity: Management skills for communication and negotiation*. Crown House Publishing, Carmarthen.

Oury, V. (2005) *5 keys to mentoring success*. www.projectsatwork.com/articles/articlesPrint.cfm?ID=223896 (accessed 18 August 2005).

Pinskey, R. (1997) *101 Ways to Promote Yourself: Tricks of the trade for taking charge of your own success*. Avon, New York.

Southgate, D. (2003) *Training grounds*. www.projectsatwork.com/article.cfm?ID=217479 (accessed 9 November 2005).

Turner, R. J. (2006) Programme and portfolio management: connecting projects to corporate strategy. *Project Manager Today*, January, 13–16.

Weaver, R. G. and Farrell, J. D. (1999) *Managers as Facilitators: A practical guide to getting work done in a changing workplace*. Berrett-Koehler, San Francisco.

Index